THERMOELECTRIC PROPERTIES OF SEMICONDUCTORS

(Proceedings of the First and Second Conferences
on Thermoelectricity)

Edited by

V. A. KUTASOV

Authorized Translation from the Russian
by A. TYBULEWICZ, B.Sc., A. Inst.P., M.I.Inf. S., F. I. L.

Springer Science+Business Media, LLC
1964

The original Russian text was published by the Academy of Sciences
Press for the Institute for Semiconductors in 1963.

ISBN 978-1-4899-5102-1   ISBN 978-1-4899-5100-7 (eBook)
DOI 10.1007/978-1-4899-5100-7

Library of Congress Catalog Card Number 64-21683

PREFACE

This collection contains the proceedings of the Second Conference on Thermoelectricity. In addition to several reviews of the physical fundamentals of thermoelectricity and of new thermoelectric materials, it includes contributions relating to experimental work on a number of materials and research methods. Some of the text deals with the technological problems of the manufacture and treatment of the materials used to make thermoelements and with the construction of thermoelectric apparatus. Also included are some papers from the First Conference on Thermoelectricity which have not been published elsewhere.

The collection is intended for the wide circle of scientists and engineers dealing with the problems of the physics of thermoelectric materials and the power applications of thermoelements.

CONTENTS

DIFFUSION IN THERMOELECTRIC MATERIALS

B. I. Boltaks and N. A. Fedorovich

The problem of stabilizing the properties of thermoelectric materials and increasing the service life of thermoelectric generators is closely related to the detailed study of the processes of impurity migration in thermoelements. These processes in thermoelements are governed by several factors, the principal ones being the following

1. The diffusion of impurities from bridging materials. This leads to a change in the contact resistance of thermoelements and as a rule causes deterioration of the device parameters.

2. Thermal diffusion due to temperature gradients in the thermoelements.

3. Electrolytic transport of matter due to the drift of ions in the electric field of thermoelements. This leads to a redistribution of impurities in thermoelements and favors the process of dissociation of complex substances.

In addition to these migration processes taking place in the interior of thermoelements, there is also the migration of particles along the free surface. Moreover, if, as is usual, thermoelements are made in the form of pressed or sintered bars, the volume migration of particles may occur both inside separate grains as well as along their boundaries. The escape and migration of impurities along grain boundaries may then be accompanied by the formation of a substance with properties completely different from those in the grain interior. In the same way, the diffusion of impurities from bridging materials may be accompanied by the formation of a new phase in the boundary layer of a thermoelement.

All these factors complicate considerably the physical picture of impurity migration in thermoelements.

In the present review we shall restrict ourselves to a narrowly-defined objective: the presentation of experimental material accumulated in recent years in the study of diffusion in thermally uniform samples of some semiconducting materials used for making thermoelectric devices. In conclusion, we shall consider briefly some general relationships governing the diffusion of impurities in semiconductors.

Experimental Data

We shall give below the results of investigations of diffusion in semiconducting materials used to prepare thermoelements. The results were mainly obtained in our laboratory and have been partly published in earlier work [1-6]. A considerable part of the data is, however, published here for the first time.

The investigations were carried out mainly using labeled atoms and the method described in [1].

1. Diffusion of Antimony and Tin in Zinc Antimonide. Zinc antimonide is one of the "original" materials in thermoelectricity and is used even now to make the positive branch of thermoelements in low-power thermoelectric generators. The diffusion of antimony and tin was investigated in coarse-grained slab samples in the temperature range from 300 to 500°C. The results obtained are shown in Fig. 1.

From the curves in Fig. 1 it may be deduced that the temperature dependence of the diffusion coefficients of antimony and tin may be expressed analytically in the form of the sum of two exponential terms:

$$D = 30 \exp\left(-\frac{1.8 \text{ eV}}{kT}\right) + 4 \cdot 10^{-11} \exp\left(-\frac{0\ 2 \text{ eV}}{kT}\right) \text{cm}^2/\text{sec}$$

for antimony, and

$$D = 2.3 \exp\left(-\frac{1.5 \text{ eV}}{kT}\right) + 3.2 \cdot 10^{-9} \exp\left(-\frac{0.375 \text{ eV}}{kT}\right) \text{cm}^2/\text{sec}$$

for tin.

The parameters given in the above formulas determine completely the diffusion coefficients of antimony and tin in SbZn at all temperatures within the region investigated. The observed kinks in the temperature dependences of the diffusion coefficients occur at the same temperature (~400°C) for both antimony and tin and are due to different diffusion mechanisms below and above this temperature. The most likely explanation is that at low temperatures (below 400°C) diffusion occurs by migration of particles along grain boundaries. For this mechanism, low activation energy values and very small pre-exponential factors are typical. Above 400°C diffusion proceeds along lattice vacancies and is characterized by high values of the activation energy and of the pre-exponential factor.

2. Self-Diffusion and Impurity Diffusion in Lead Telluride.

The self-diffusion of lead and tellurium, as well as diffusion of antimony and tin, in PbTe was investigated using single-crystal p-type ingots with carrier densities of approximately 10^{17}cm^{-3}. The diffusion of lead was investigated in the range 250-500°C using the p − n junction method. The diffusion of other elements was studied in the range 500-800°C by means of labeled atoms. The results obtained for the diffusion coefficients at different temperatures are given in Fig. 2, and the diffusion parameters calculated from these results are given below:

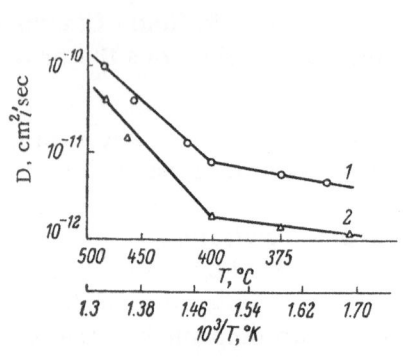

Fig. 1. Temperature dependence of the diffusion coefficients of tin and antimony in SbZn. 1) Sn; 2) Sb.

Element	$\Delta\varepsilon$, eV	D_0, cm^2/sec
Pb	0.6	$2.9 \cdot 10^{-5}$
Te	0.75	$2.7 \cdot 10^{-6}$
Sb	1.54	$4.9 \cdot 10^{-2}$
Sn	1.56	$3.1 \cdot 10^{-2}$

From these values it is clear that the diffusion parameters of lead and tellurium, on the one hand, and of antimony and tin, on the other, are quite similar. However, the activation energies for self-diffusion are considerably smaller than for the diffusion of foreign atoms (Sb and Sn). The higher values of the activation energy in the latter case are probably related to the formation of molecular complexes of the type [SnTe] and [SbTe], for which the binding energies are greater than the binding energy of PbTe.

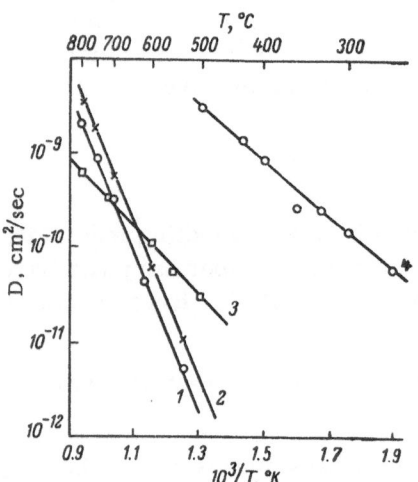

Fig. 2. Self-diffusion and diffusion of impurities in PbTe. 1) Sn; 2) Sb; 3) Te; 4) Pb.

3. Diffusion in Lead Selenide.

Up to now, data on diffusion in lead selenide have been obtained for only two elements: selenium and antimony. Measurements of the

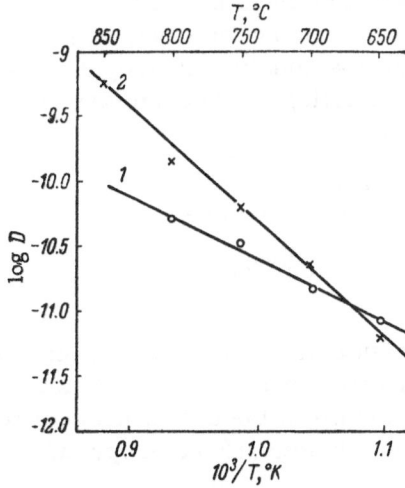

Fig. 3. Diffusion of selenium and antimony in PbSe. 1) Se; 2) Sb.

diffusion coefficients carried out on single-crystal samples at various temperatures in the range 650-850°C gave the results shown in Fig. 3 and can be expressed by the following formulas:

$$D_{Se \to PbS} = 2.1 \cdot 10^{-5} \exp\left(-\frac{1.2 eV}{kT}\right) \; cm^2/sec,$$

$$D_{Sb \to PbSe} = 0.34 \exp\left(-\frac{2.0 \, eV}{kT}\right) \; cm^2/sec.$$

As in the case of PbTe, the activation energy for self-diffusion (Se) is smaller than the activation energy for a foreign element (Sb). It is suggested that again complexes of the type [SbSe] are formed.

The considerably greater activation energy for self-diffusion and the diffusion of impurities in PbSe compared with PbTe is in agreement with the difference between the heats of formation of these two compounds.

4. Diffusion of Impurities in Bismuth Telluride. Bismuth telluride Bi_2Te_3 and solid solutions based on it ($Bi_2Te_3 - Bi_2Se_3$ and $Bi_2Te_3 - Sb_2Te_3$) are widely used at present for the manufacture of thermoelectric devices. This was the reason for investigating impurity diffusion processes in these materials.

Bismuth telluride (like bismuth selenide) has a complex structure (Fig. 4) with a strong anisotropy of the principal physico-chemical properties. Therefore it was expected that the diffusion coefficients of various elements in Bi_2Te_3 would also depend on the crystallographic direction.

Figure 5 gives the results obtained in a study of the diffusion of silver along and at right angles to the cleavage planes of Bi_2Te_3 crystals. This figure shows that at 500°C the diffusion coefficients in these two mutually perpendicular directions differ by a factor of almost 1000, and at 100°C this difference reaches eight orders of magnitude. The variation of the diffusion coefficients with temperature may be expressed by the following relationships:

$$D_{\parallel Ag \to Bi_2Te_3} = 2.2 \cdot 10^{-3} \exp\left(-\frac{0.42 \, eV}{kT}\right) \; cm^2/sec,$$

$$D_{\perp Ag \to Bi_2Te_3} = 2.3 \cdot 10^{-1} \exp\left(-\frac{1.17 \, eV}{kT}\right) \; cm^2/sec.$$

Carlson [7] investigated the diffusion of copper in bismuth telluride. The results of this investigation (Fig. 6) show that copper also diffuses very rapidly along cleavage planes, even at room temperature. For this direction the value of the diffusion coefficient of copper at room temperature reaches about 10^{-6} cm²/sec, which is several orders of magnitude greater than any other known diffusion coefficient of an element in a solid. At right-angles to the cleavage planes the diffusion coefficient of copper is considerably smaller.

The temperature dependences of the diffusion coefficients of copper for the two mutually perpendicular directions are, according to Carlson [7], given by the following relationships:

$$D_{\parallel Cu \to Bi_2Te_3} = 3.4 \cdot 10^{-3} \exp\left(-\frac{0.2 \, eV}{kT}\right) \; cm^2/sec,$$

$$D_{\perp Cu \to Bi_2Te_3} = 7.1 \cdot 10^{-2} \exp\left(-\frac{0.8 \, eV}{kT}\right) \; cm^2/sec.$$

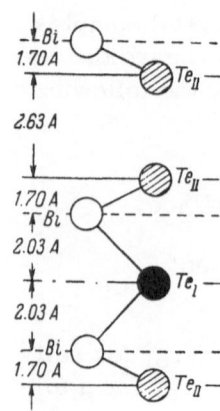

Fig. 4. Structure of bismuth telluride.

We have also investigated the diffusion of cadmium in Bi_2Te_3. The results obtained are given in Fig. 7. As in the case of silver and copper, the rate of diffusion of cadmium in Bi_2Te_3 is strongly anisotropic. The temperature dependences of the diffusion coefficients of cadmium are given by the following relationships:

$$D_{\parallel Cd \to Bi_2Te_3} = 4.8 \cdot 10^{-3} \exp\left(-\frac{0.48 \text{ eV}}{kT}\right) \text{ cm}^2/\text{sec},$$

$$D_{\perp Cd \to Bi_2Te_3} = 10^2 \exp\left(-\frac{1.8 \text{ eV}}{kT}\right) \text{ cm}^2/\text{sec}.$$

The table lists the diffusion coefficients of copper, silver, and cadmium for the two mutually perpendicular directions in Bi_2Te_3 at the same temperature of 500°C. The table also gives the activation energy values $\Delta\varepsilon$, pre-exponential factor D_0, and the activation entropy ΔS.

The table shows that the diffusion coefficient parallel to the cleavage planes is large for all these elements. There are grounds for assuming that the diffusion along this direction proceeds through interstices along layers of like atoms, and that it is not related to the considerable departures from order in the Bi_2Te_3 lattice. This is indicated by

Diffusion Coefficients of Copper, Silver, and Cadmium

Element	Ionic radius, A	Parallel diffusion				Perpendicular diffusion			
		D_{500} °C cm²/sec	$\Delta\varepsilon$ eV	D_0 cm²/sec	ΔS	D_{500} °C cm²/sec	$\Delta\varepsilon$ eV	D_0 cm²/sec	ΔS
Cu . . .	0.96	$1.6 \cdot 10^{-4}$	0.20	$3.4 \cdot 10^{-3}$	0	$1 \cdot 10^{-6}$	0.8	$7.1 \cdot 10^{-2}$	k
Ag . . .	1.13	$1.6 \cdot 10^{-6}$	0.42	$2.2 \cdot 10^{-3}$	0	$1 \cdot 10^{-8}$	1.17	$2.3 \cdot 10^{-1}$	$6.4\,k$
Cd . . .	0.97	$6.3 \cdot 10^{-7}$	0.48	$4.8 \cdot 10^{-3}$	k	$2 \cdot 10^{-10}$	1.80	100	$6.8\,k$

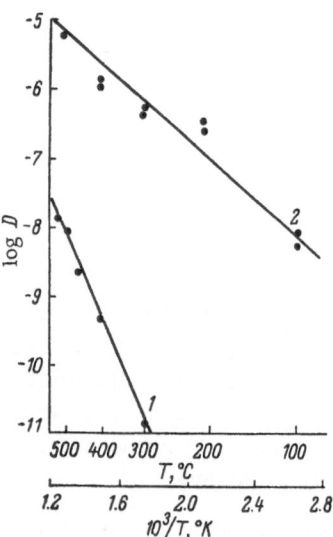

Fig. 5. Anisotropy of the diffusion of silver in bismuth telluride. 1) $D_\perp = 2.3 \cdot 10^{-1} \exp(-1.17/kT)$; 2) $D_\parallel = 2.2 \cdot 10^{-3} \exp(-0.42/kT)$.

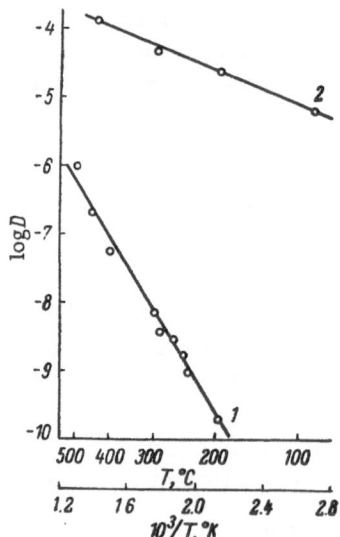

Fig. 6. Anisotropy of the diffusion of copper in bismuth telluride. 1) $D_\perp = 7.1 \cdot 10^{-2} \exp(-0.8/kT)$; 2) $D_\parallel = 3.4 \cdot 10^{-3} \exp(-0.2/kT)$.

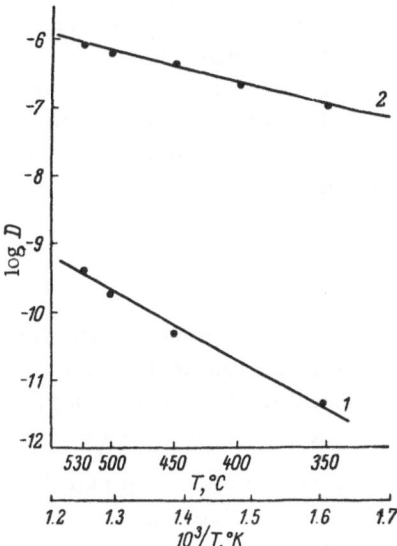

Fig. 7. Anisotropy of the diffusion of cadmium in bismuth telluride. 1) $D_\perp$ = $10^{-2} \exp(-1.8/kT)$; 2) $D_\parallel$ = 4.8 · $10^{-3} \exp(-0.48/kT)$.

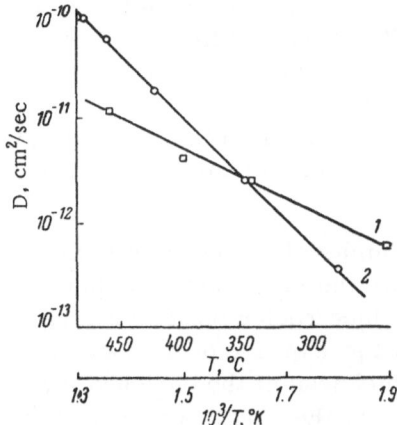

Fig. 8. Diffusion of tin and antimony in Bi_2Te_3. 1) Sn; 2) Sb[$D_{Sb \to Bi_2Te_3}$ = 4.3 · $10^{-4} \exp(-1.04/kT)$; $D_{Sn \to Bi_2Te_3}$ = 3 · $10^{-8} \exp(-0.5/kT)$].

the low values of the activation entropy ΔS. Both in the case of copper and silver, $\Delta S = 0$. For cadmium, $\Delta S \simeq k$.

Diffusion at right angles to the cleavage planes may occur either along interstices or along bismuth vacancies. In the former case the activation energy should increase with increase of the electronegativity of the diffusing element, because this favors formation of ionic bonds in the Bi_2Te_3 lattice. In the case of diffusion along vacancies, the binding energy should increase with increase in the number of valence electrons of the diffusing elements and decrease with increase of the difference between the covalent radii of copper, silver, and cadmium on the one hand, and the radius of bismuth on the other.

In addition to studies of the diffusion of elements of the first and second groups (Cu, Ag, and Cd), an investigation has been made of the diffusion of tin and antimony in Bi_2Te_3 [8].

The diffusion coefficients were measured in the temperature range 260-500°C on imperfect crystals at right angles to the cleavage planes. The results for the diffusion of tin and antimony are shown in Fig. 8, and they are satisfactorily described by the following relationships:

$$D_{Sn \to Bi_2Te_3} = 3.0 \cdot 10^{-8} \exp\left(-\frac{0.5 \text{ eV}}{kT}\right) \text{ cm}^2/\text{sec},$$

$$D_{Sb \to Bi_2Te_3} = 4.3 \cdot 10^{-4} \exp\left(-\frac{1.08 \text{ eV}}{kT}\right) \text{ cm}^2/\text{sec}.$$

It is worth noting that similar measurements of the diffusion coefficients of tin and antimony have also been carried out on bismuth selenide. In that case the results may be described by similar relationships:

$$D = 4 \cdot 10^{-4} \exp\left(-\frac{0.41 \text{ eV}}{kT}\right) \text{ cm}^2/\text{sec},$$

for the diffusion of tin, and

$$D = 1.8 \cdot 10^{-3} \exp\left(-\frac{1.27 \text{ eV}}{kT}\right) \text{ cm}^2/\text{sec},$$

for the diffusion of antimony.

From these results it is clear that both in bismuth telluride and in bismuth selenide the diffusion of antimony is represented by a considerably higher activation energy than the diffusion of tin. There are reasons for assuming that in these two compounds antimony diffuses along bismuth vacancies, forming complexes of the type [SbTe] and [SbSe] with considerably higher binding energies than the binding energy of atoms in the host lattice. This is not so for tin, because compounds of tin with tellurium and selenium have different crystal structure than bismuth telluride and bismuth selenide. The formation of complexes of tin with tellurium or selenium is unlikely (such complexes are unstable) since this would involve strong deformation of the lattice. Tin may diffuse in bismuth telluride and selenide by motion along interstices or along bismuth vacancies, becoming bound only weakly to the neighboring atoms in the layer.

5. **Diffusion of Impurities in the Solid Solutions $Bi_2Te_3 - Bi_2Se_3$ and $Bi_2Te_3 - Sb_2Te_3$.** We have carried out a series of diffusion studies on polycrystals of these solid solutions, which are used in the manufacture of thermoelements. First, we investigated the diffusion of antimony in a solid solution of composition $(25\% \, Bi_2Te_3 + 75\% \, Sb_2Te_3) + 3\% \, Te$, which is used to make the positive branch of thermocouples. The study was carried out on samples cut from ingots prepared by the method of directional casting, in the temperature range 300-500°C. The experimental results obtained are described by the following dependence:

$$D = 2.5 \cdot 10^{-7} \exp\left(-\frac{0.51 \text{ eV}}{kT}\right) \text{ cm2/sec.}$$

Comparision of the diffusion parameters obtained in this way with the parameters for the diffusion of antimony in Bi_2Te_3 shows that the activation energy for the diffusion of Sb in the ternary alloy is considerably lower than in bismuth telluride. It is possible that in the alloy there is a considerable contribution from the diffusion along many of the grain boundaries.

Next we investigated the diffusion of silver in $Bi_2Te_3 - Bi_2Se_3$, which is used to make the negative branch of thermocouples. It was found that the rate of diffusion of silver in this solid solution is considerably greater than even the rate of diffusion of silver in Bi_2Te_3 crystals along the cleavage planes.

Using radioactive isotopes we investigated the redistribution of silver between the interior and surface of the sample. This study was carried out on hot-pressed samples of $Bi_2Te_3 - Bi_2Se_3$. Radioactive silver Ag^{110} was introduced into the melt of the components in an evacuated quartz ampoule, and special attention was paid to the homogenization of the alloy, which was achieved by frequent mixing of the melt. The uniformity of the distribution of the silver was checked by successive sectioning of the samples and measuring the radioactivity of each section. The samples were then annealed, and again the distribution of silver was determined. The annealing was carried out in various gaseous media: in air, in argon, in hydrogen, and in vacuum ($\sim 10^{-3}$ mm Hg). The annealing temperature was 300°C and its duration varied from 3 to 120 hr.

Figure 9 gives the distribution of silver with depth in the sample after annealing in air for 3 hr. The figure shows that this annealing altered considerably the initial uniform distribution of silver. The surface region became strongly enriched with silver at the expense of its content in the interior. The thickness of the enriched layer increased with increase of the duration of annealing, and the concentration minimum shown in Fig. 9 shifted deeper into the samples.

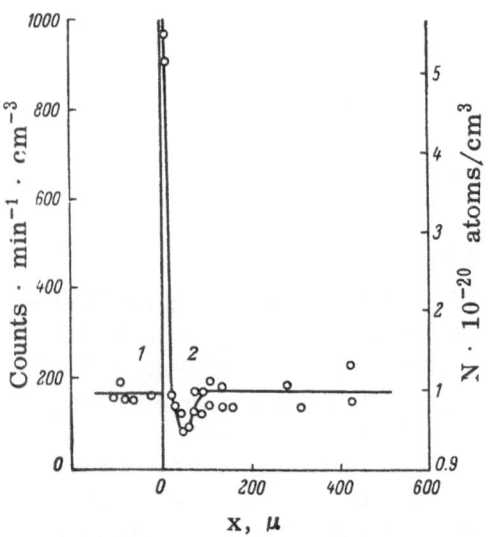

Fig. 9. Distribution of silver in a sample on annealing in air for 3 hr. 1) Before annealing; 2) after annealing.

This is not observed on annealing in a reducing atmosphere, while on annealing in unpurified argon (containing traces of oxygen) and in vacuum there is only a slight redistribution of silver between the interior and the surface layers.

This investigation thus shows unambiguously that the ambient gaseous medium plays an important role in the process of migration of silver from the interior to the surface, and that in an oxidizing medium this process occurs much more rapidly. As the sample is oxidized and the oxide layer grows, the escape of silver to the surface becomes more and more rapid; at the surface it is dissolved and bound chemically to the oxide. The silver itself may act as an accelerator of the oxidation process, since its emergence on the surface is accompanied by an increase of the free-electron density necessary for the oxidation of the alloy. Indeed, experiment shows that in the absence of silver, the oxidation of $Bi_2Te_3 - Bi_2Se_3$ samples proceeds very slowly.

To stabilize the parameters of thermoelements made from the solid solution $Bi_2Te_3 - Bi_2Se_3$, one uses as alloying admixtures not the elements of group I but their halides: CuBr, AgI, and others. In connection with this, we carried out several experiments using radioactive isotopes in order to find the nature of the redistribution of silver in pressed $Bi_2Te_3 - Bi_2Se_3$ samples containing iodine (iodine was introduced into the alloy in the form of TeI_4).

These experiments showed there was no marked redistribution of silver in the presence of iodine. Similar results were obtained on annealing the samples in an atmosphere of hydrogen.

The results obtained show that iodine, by combining with silver, prevents its emergence on the surface. It is reasonable to expect similar behavior for many other impurities in the presence of halogens.

On Certain Factors Which Affect the Diffusion Coefficients of Impurities in Semiconductors

The diffusion of impurities in semiconducting materials is accompanied by a complex process of interaction of the diffusing particles with the neighboring lattice atoms (ions), structure defects, impurities, and also with one another when their concentration is high. Recent experimental results on diffusion, mainly in germanium and silicon, allow us to draw conclusions about the factors which affect the diffusion coefficients of impurities in semiconductors [1]. It seems useful to list here some of the factors which must be allowed for in the problem of the stabilization of the properties of thermoelectric materials and of thermoelement parameters.

1. In the case of impurities forming substitutional solid solutions in semiconductors with covalent bonds, the main factor that governs the rate of diffusion is the number of valence electrons in the diffusing particle. Impurity atoms with valence electrons insufficient in number to form complete covalent bonds with neighboring lattice atoms always move rapidly, and their rate of diffusion increases as the number of electrons necessary for completion of the bonds increases.

In semiconductors with ionic binding the governing factors also include the electron affinity and the value of the electronegativity of the diffusing particle. The rate of diffusion in this case increases with the difference between the number of valence electrons, electron affinity, and electronegativity of the diffusing particle, and the corresponding properties of that component of the ionic compound whose vacancies are used in the diffusion.

2. The rate of diffusion also depends on the atomic (ionic, covalent) radius of the diffusing particle. The larger the difference between the radius of the diffusing particle and the radii of the particles in the host compound, the greater the distortions in the host lattice and the slower the diffusion.

3. The presence of structure defects increases the rate of diffusion, especially when the migration takes place by vacancy jumps.

4. The rate of diffusion in semiconductors is affected considerably by the Coulomb interaction between the diffusing particles and vacancies in the host lattice. When the charges of the diffusing particle and the nearby vacancy are of like sign, the Coulomb interaction increases the activation energy and consequently reduces the diffusion rate. When the charges are of opposite sign the Coulomb interaction accelerates the rate of diffusion.

5. A considerable influence on the rate of diffusion in semiconductors is also exerted by impurities. The role of impurities, considered here as the third component, is very varied.

The presence of impurities produces local distortions in the crystal lattice and this accelerates the rate of migration of both the host and foreign particles. Moreover, in the presence of charged impurity particles there is an additional electrostatic interaction between the diffusing particles and the uniformly distributed other ionized impurities.

The presence of a uniformly distributed acceptor impurity slows down the rate of diffusion of donors and accelerates the diffusion of other acceptors. Similarly, a uniformly distributed donor impurity reduces the rate of diffusion of acceptors and increases the corresponding rate of other donors.

The presence of impurities may also lead to the formation of complexes which reduce strongly the rate of diffusion, particularly when the migration of the diffusing particles proceeds by successive interstitial jumps. The results on the influence of iodine on the diffusion redistribution of silver in the solid solution $Bi_2Te_3 - Bi_2Se_3$ give an example of this.

6. In semiconductors, as in metals, there is a definite correlation between the rate of diffusion and the solubility of foreign particles. The factors enhancing the solubility slow down the rate of diffusion, and, conversely, the factors reducing the solubility accelerate the diffusion.

The list given here does not cover all the possible interactions between diffusing particles in the crystal lattice of a semiconductor. We have indicated only the factors affecting the diffusion coefficient, when the diffusion is not accompanied by phase transformations. A phase transition in the case of reactive diffusion contributes additional factors which affect the diffusion processes, but this is outside the scope of the present article.

Conclusions

In the present communication we have given the experimental data and individual results on the diffusion of impurities in some semiconducting materials used in the manufacture of thermoelements. Successful solution of the problems involved in the making of thermoelements is undoubtedly related to the further extension of studies of the diffusion processes occurring in thermoelectric materials and in thermoelements made from these materials. These studies should be developed both in the direction of the accumulation of additional data on the diffusion coefficients and migration mechanism of various impurities in thermoelectric materials, as well as in the direction of the investigation of diffusion processes in thermoelements under working conditions. In this connection, special attention should be paid to the factors which influence the rate of migration of impurities along crystal boundaries. These processes are more complex than the volume diffusion in crystals, since they are frequently accompanied by the formation of new phases.

An important problem is the investigation of the possibility of slowing down the rate of diffusion in thermoelectric materials. One of the methods of such control of the rate of diffusion is as pointed out above, the introduction of a third component. However, this also alters the electrical properties of the material. Therefore, the diffusion studies should be accompanied by an investigation of the influence of the additional impurity on the electric properties.

It is also necessary to pay attention to the processes of impurity migration in the electric field in thermoelements and to the thermal diffusion due to the presence of a temperature drop in thermoelements. These phenomena play an important role in the processes of impurity distribution and change the electrical properties of thermoelements.

We have pointed out here only some of the more important problems connected with the study of the diffusion in thermoelectric materials and thermoelements made from them. A detailed and comprehensive study of the diffusion processes in these materials and devices requires great and concerted efforts from both physicists and chemists.

LITERATURE CITED

1. B. I. Boltaks, Diffusion in Semiconductors, Fizmatgiz, 1961; English translation: Infosearch Ltd., London, 1963.
2. C. Wert and C. Zener, Phys. Rev. 76: 1169, 1949.
3. B. I. Boltaks and G. S. Kulikov, Zhur. Tekh. Fiz. 27: 82, 1957.
4. E. Brady, J. Electrochem. Soc. 101: 466, 1954.
5. B. I. Boltaks and Yu. N. Mokhov, Zhur. Tekh. Fiz. 26: 2448, 1956.
6. B. I. Boltaks and Yu. N. Mokhov, Zhur. Tekh. Fiz. 28: 1046, 1958.
7. R. O. Carlson, J. Phys. Chem. Solids 13: 65, 1960.
8. B. I. Boltaks, Zhur. Tekh. Fiz. 25: 767, 1955.

ENERGY SPECTRUM OF CARRIERS
IN THERMOELECTRIC MATERIALS

Yu. V. Ilisavskii

All charge transport processes in semiconductors, including thermoelectric emf and electrical conductivity, depend on the form of the carrier energy spectrum.

Group theory gives several variants of band structure which do not contradict the symmetry of a particular crystal. Which of these variants applies to a given substance is established by experimental studies of cyclotron resonance [1, 2], magnetoresistance [1], and the effect of hydrostatic pressure and uniaxial deformation (piezoresistance effect) on the electrical and optical properties of semiconductors [3-11].

Recently, important information on the band structure of some semiconductors has been obtained by combining hydrostatic pressure with uniaxial deformation [12] and by investigating the de Haas-van Alphen effect [13].

Bands in Cubic Crystals

With the exception of Bi_2Te_3 and Bi_2Se_3, all the materials and alloys considered here (Ge and Si, PbTe, PbSe, PbS) have cubic structure. From symmetry considerations it follows that in cubic crystals the following variants of the energy minima distributions are possible: 1) a minimum in the center of the Brillouin zone; the band is then either simply spherical (if the energy levels are not degenerate), or (in the case of level degeneracy) it consists of a set of several energy surfaces joined at the point $\mathbf{k} = 0$; 2) a minimum not in the center of the Brillouin zone ($\mathbf{k} \neq 0$). Applying the symmetry transformations to this point we obtain a network of equivalent minima, i.e., a band with many minima (a many-valley band). The number of minima in the band is independent of the position of the original minimum. If the latter lies on the [100] symmetry axis and inside the Brillouin zone, we find six equivalent minima with centers along the [100] axes; a minimum on the [111] axis gives rise to eight equivalent minima, while a minimum on the [110] axis gives 12 minima. If the original minimum lies on the Brillouin zone boundary, then the number of the equivalent minima is halved, because each of these minima belongs to two Brillouin zones. Since the [100] and [111] axes are, respectively, a fourfold and a threefold symmetry axis, the minima of the <100> and <111> bands are ellipsoids of revolution. The $\{110\}$ axes are twofold axes and therefore the energy surfaces of the <110> band are ellipsoids.

If the original minimum does not lie on any symmetry element, the number of equivalent minima may be very large (48 for the symmetry class m3m). Moreover, as in the case of a minimum in the center of the Brillouin zone, the energy levels of such bands may be degenerate. However, up to now such complex bands have not been observed in studies of the carrier energy spectrum in crystals, and we shall therefore restrict ourselves to considering the <000>, <100>, <111>, and <100> bands.

It is convenient to begin a discussion of the carrier energy spectrum in solid solutions of the Ge − Si system by considering the band structure of the initial components Ge and Si.

Germanium. Theoretical calculations [14] and investigations of magnetoresistance and piezoresistance [1, 3, 4], of cyclotron resonance [1, 2], and of optical and magnetic effects [15, 16] have established the dependence of the carrier energy in Ge on the magnitude and direction of the wave vector **k** [17, 18]. Figure 1 shows this dependence for two directions: [100], $k_{max} = (2\pi/a_0, 0, 0)$ and [111], $k_{max} = (\pi/a_0, \pi/a_0, \pi/a_0)$.

It is clear from Fig. 1 that the conduction band minimum lies along the [111] direction on the Brillouin zone boundary while the <000> and <100> bands lie above, separated by about 0.18 eV. Energy surfaces of the <111> band are ellipsoids of revolution, elongated along the {111} symmetry axes. There are four ellipsoids in the band. The ratio of the longitudinal and transverse effective masses is $m_{||}/m_\perp \simeq 20$ and it depends weakly on temperature [19]. The transverse effective mass is $m_\perp \simeq 0.08 m_0$.

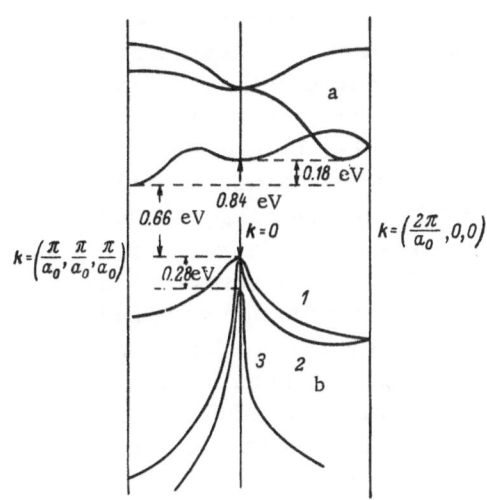

Fig. 1. Variation of E(**k**) for Ge along the directions [100] and [111]. a) Conduction band; b) valence band.

The valence band of germanium consists of three energy surfaces with maxima at the point **k** = 0. Because of this the valence band is triply degenerate (without allowance for spin), but because of the spin-orbit interaction the lower band 3 is displaced downwards by $\Delta \simeq 0.28$ eV. The bands 1 and 2 have different curvature at the point **k** = 0, i.e., different effective masses (light and heavy holes: $m_l \simeq 0.04 m_0$, $m_h \simeq 0.3 m_0$ [18]). The cross sections of the constant energy surfaces of these bands by the plane (100) have the form shown in Fig. 2.

Silicon [1, 2, 14, 15, 17, 18]. The valence band of silicon differs from the valence band of germanium only by different values of the effective masses of light and heavy holes ($m_l \simeq 0.17 m_0$, $m_h \simeq 0.5 m_0$) and of the spin-orbit splitting constant ($\Delta \simeq 0.05 m_0$ [18]), as shown in Fig. 3.

In the conduction band of silicon the <100> minima lie below all the other minima (Fig. 3). The <000> and <111> bands lie about 1.5 eV above the <100> band. The energy surfaces are ellipsoids of revolution elongated along the {100} symmetry axes. Since the minima are inside the Brillouin zone, there are six ellipsoids. The anisotropy of the ellipsoids is less than in the case of germanium ($m_{||}/m_\perp \simeq 5.2$), and the effective electron mass is greater ($m_\perp \simeq 0.19 m_0$).

Ge – Si Solid Solutions. The conduction bands of Ge and Si differ markedly, and therefore on formation of Ge–Si solid solutions and transition from one component to the other there should be a change of the band structure. This is so in practice, and the dependence of the forbidden-band width on the solid-solution composition [20-22] has, according to electrical and optical measurements, a sharp discontinuity at about 14-15% Si [22]. Direct information on the change in the band structure should be available from cyclotron resonance. Unfortunately the relaxation time decreases considerably in solid solutions with a high content of the second component, and therefore cyclotron resonance can be observed only in solid solutions containing a predominance of one component. Such cyclotron resonance measurements have been carried out [23] and they have shown that when the content of the second component is low (0.4 or 0.75% Ge in Si; 0.8 or 5.4% Si in Ge) the band structure is not greatly affected. The piezoresistances of germanium and silicon differ considerably as regards anisotropy, and therefore investigation of the piezoresistance of single-

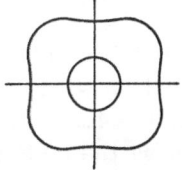

Fig. 2. Cross section of the energy surfaces of the Ge valence band by a (100) plane near the point **k** = 0.

11

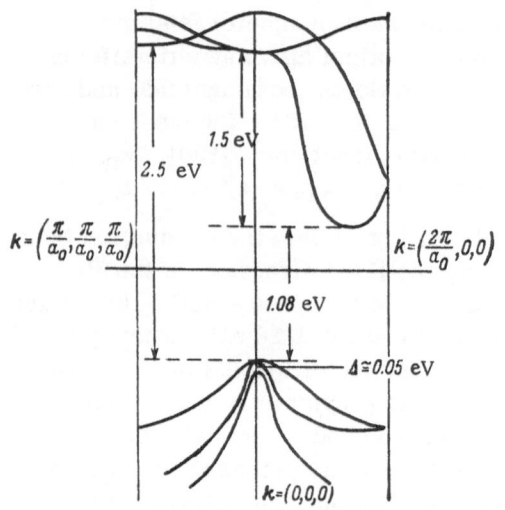

$k = \left(\frac{\pi}{a_0}, \frac{\pi}{a_0}, \frac{\pi}{a_0}\right)$

1.5 eV

2.5 eV

$k = \left(\frac{2\pi}{a_0}, 0, 0\right)$

1.08 eV

$\Delta \approx 0.05$ eV

$k = (0,0,0)$

Fig. 3. Variation of $E(\mathbf{k})$ for Si along the directions [100] and [110].

crystal samples of Ge−Si alloys is of great interest. However, up to now such piezoresistance measurements have not been carried out, although a method of preparing single-crystal samples of Ge−Si alloys with up to 23% Si has been developed [26]. The answer to the question of whether there is a change of the band structure in the Ge−Si system has been obtained from investigations of the magnetoresistance effect [24-26]. It has been found that the anisotropy and distribution of <111> ellipsoids in the conduction band of germanium remain practically unaffected up to compositions approaching 7% Si in Ge. The anisotropy of the ellipsoids has then been found to decrease and in the region of 11-14% Si in Ge a transition from the <111> band to the <100> band has been observed. All these data lead to the conclusion [27] that on transition from Ge to Si the <111> band is displaced upward and the <100> band is displaced downward, until at Si concentrations close to 14 % 'the two bands occupy the same position. The results obtained lead to an interesting and not hitherto obvious conclusion: the anisotropy of energy minima of <100> and <111> in Ge and Si is the same, but in the Ge the "germanium" <111> minimum is lower while in Si it is the "silicon" <100> minimum that is lower.

This conclusion, and the assumption about the nature of motion of the bands of various types when the composition of the Ge−Si solid solution is varied, have been confirmed by experiments on hydrostatic compression [28]. It has been found that on increase of the hydrostatic pressure, and consequent reduction of the lattice constant a_0 (the transition from Ge to Si also involves a reduction of the lattice constant), the forbidden-band width of Ge first increases and then decreases, while the forbidden-band width of silicon decreases monotonically with increasing pressure. Later measurements [8, 29] have shown that on the reduction of the distance between atoms the <111> band of Ge is displaced upward and the <100> band moves downward. At $P = 5 \cdot 10^4$ atm, these two bands are at approximately the same level, and then the <100> minima move downward. The anisotropy of these minima is similar to the anisotropy of the <100> band of Si.

Lead Chalcogenides (PbTe, PbSe, PbS). Lead chalcogenides are also cubic (NaCl-type lattice), and therefore we can expect the same variants of the energy spectrum as in Ge and Si. A high carrier density which remains constant down to very low temperatures is a characteristic feature of all compounds in the PbS group. This prevents the observation of cyclotron resonance, and therefore other methods have to be used to investigate the band structure of PbTe, PbSe, and PbS.

Lead Telluride. Investigations of the magnetoresistance [30-34] and piezoresistance [5, 35, 36] of single-crystal samples of p-type PbTe have shown that the main contribution to transport processes comes from the ellipsoids of revolution elongated along the {111} symmetry axes. According to Allgaier [33], the anisotropy of mobilities in an ellipsoid of the valence band is $u_\perp / u_\parallel = m_\parallel \tau_\perp / m_\perp \tau_\parallel = 4.74$ at T = 293°K.

These results have been confirmed by studies of the de Haas-van Alphen effect [13, 37]; it has been found that the valence band of PbTe consists of four ellipsoids of revolution with centers on the Brillouin zone boundaries and elongated along the {111} axes. At T = 4.2°K the effective mass ratio is $m_\parallel / m_\perp = 6.4$, and $m_\perp = 0.04 m_0$. However, in investigating the de Haas-van Alphen effect it has been found that, as well as the <111> band, there is also a spherical band with its center at the point $\mathbf{k} = 0$, and with an effective density-of-states mass ($m_D^* = 0.12 m_0$)

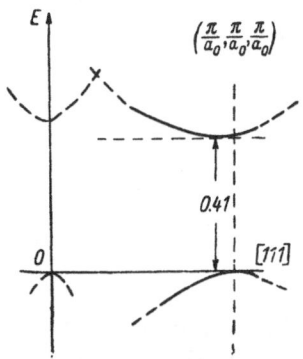

Fig. 4. Positions of the energy minima in PbTe.

close to the effective density-of-states mass of the <111> band. At T = 4.2°K the edges of both bands have practically the same energy ($\triangle E = 0.002 \pm 0.002$ eV). This very interesting result requires verification, since measurements at higher temperatures have not yet shown the presence of a spherical band similar to the <111> band.

The conduction band of PbTe, as indicated by the magneto-resistance and piezoresistance data [5, 33, 34, 38], is similar to the valence band with the difference that the anisotropy of the el-lipsoids of revolution is somewhat smaller: $m_\| / m_\perp = 4.5$, $m_\perp = 0.025 m_0$ at T = 4.2°K [38], and $u_\perp / u_\| = 3.3$ at 293°K [34]. Data on the oscillations of magnetoresistance [38] seem to suggest that the conduction band of PbTe includes a spherical band similar to the <111> band.

Figure 4 gives the distribution of minima in PbTe. Since the positions of these spherical bands are not yet finally established, the corresponding levels are represented by dashed curves

Investigation of the piezoresistance effect has made it possible to establish an interesting property of the band structure of PbTe. The strong piezoresistance of semiconductors is usually related to the fact that energy minima which are not equivalent are affected under uniaxial deformation in different ways. Some carriers are transferred from the minima with higher energies to the minima with lower energies (the overflow effect); the anisotropy of the latter becomes predominant and this alters the resistivity. The anisotropy of piezoresistance is closely related to the band anisotropy, and the effect is proportional to 1/T [6].

Deformation, by changing the interatomic distances and bond angles, may also alter the form of the energy surfaces (i.e., the effective mass) as well as their positions [39]. The change in the effective mass on deformation also leads to a change of the resistance. It is important to note that this contribution to the piezoresistance effect depends weakly on temperature [39]. In Ge and Si the change of the effective mass on deformation is slight. In PbTe this change is considerably greater. Since temperature variation also alters the average interatomic distances, the effective mass in PbTe should depend on temperature, which is confirmed by experiment [40].

Lead Selenide. There is as yet very little information on the band structure of PbSe. Preliminary results of measurements of the magnetoresistance of n- and p-type samples of PbSe at room temperature [33, 34] indicate weak anisotropy of the energy surfaces. Investiga-

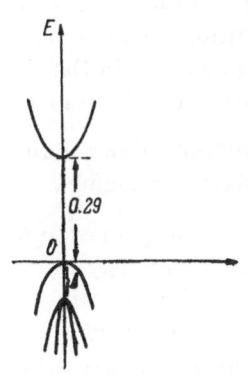

Fig. 5. Band model of PbSe.

tion of the piezoresistance of p-type PbSe single crystals at 78-300°K [5] has shown that the effect is independent of temperature and that the main cause of the resistance change on deformation is the change of the effective mass. * Thus the available data suggest that the energy minima are situated in the center of the Brillouin zone. Consequently the results obtained from the piezoresistance effect may be accounted for by the PbSe band model proposed by Bir and Pikus [41] (Fig. 5).

It is assumed that the conduction band is simply spherical and the valence band (without allowance for spin) is triply degenerate. As in p-type Ge and p-type Si, the spin-orbit interaction splits the energy levels of the valence band and partly removes the degeneracy. However, according to Bir and Pikus, in contrast to p-type Ge and p-type Si, the two lower bands are split off together. The spin-orbit splitting Δ may reach 1 eV (because the average atomic number of PbSe is 58).

Consequently, as in PbTe, m depends on temperature [9].

Comparision of the Energy Spectra of n- and p-type Bi_2Te_3 (T = 78°K) [47]

Anisotropy parameters	n-Type		p-Type	
	solution I	solution II	solution I	solution II
w_1	10.96	0.0565	11.73	0.0603
w_2	0.817	0.236	1.02	0.277
$\cos^2 \theta$	0.945	0.540	0.921	0.560

Another possible variant which agrees with the experimental data (almost spherical isotropic "ellipsoids" with centers at the point $\mathbf{k} = 0$) is unlikely to be correct. On the one hand, the condition $u_\perp / u_\parallel = 1$ imposes rigid restrictions on the values of a large number of independent constants [6]. On the other hand, as indicated by the dependence of the effective mass on the interatomic distance (the width of the forbidden band), the matrix element of the interaction between the conduction and valence bands is not equal to zero, and this should produce a considerable anisotropy in the energy minima.

Lead Sulfide. The results of measurements of the magnetoresistance of n-type and p-type PbS at room temperature [33, 34] indicate that the anisotropy of the energy surfaces in PbS is small. Investigations of the de Haas-van Alphen effect in n-type PbS at T = 4.2°K [13] also failed to show anisotropy of the energy surfaces greater than the experimental error. To explain quantitatively the results obtained, it is necessary to assume that the conduction band consists either of < 111 > minima with centers on the Brillouin zone boundaries and $u_\perp / u_\parallel \simeq 1$ (almost spherical structure), or of two bands with low anisotropy.

Measurements in the temperature range 78-300°K [42] indicated that the piezoresistance of n-type PbS is low (in agreement with the data on the magnetoresistance and the de Haas-van Alphen effect), being governed to a considerable extent by the change in the effective mass. However, we can separate from the piezoresistance a part which is proportional to 1/T with an anisotropy corresponding to the < 111 > band or to a degenerate band with its center at the point $\mathbf{k} = 0$. The cause of this may be the close position of the two types of band. Since the available data do not permit an unambiguous interpretation, it is difficult to suggest a sufficiently reliable band structure for PbS.

Bismuth Chalcogenides (Bi_2Te_3, Bi_2Se_3). These two compounds crystallize with rhombohedral structure (symmetry class $R\overline{3}m$). Therefore Bi_2Te_3 and Bi_2Se_3 crystals have the following symmetry elements: a threefold axis C_3, a plane of symmetry parallel to C_3, and a twofold axis C_2 perpendicular to the symmetry plane. In crystals with such symmetry elements the following energy spectrum variants are possible [43]: 1) one ellipsoid of revolution at the point $\mathbf{k} = 0$; 2) two ellipsoids of revolution on the axis C_3; 3) six ellipsoids on the axes C_2; 4) six ellipsoids with centers in the planes of symmetry; 5) 12 ellipsoids not lying on the symmetry elements.

Since these two chalcogenides retain high carrier density down to liquid helium temperatures, it is difficult to observe cyclotron resonance.

Bismuth Telluride. Investigation of the galvanomagnetic properties of n- and p-type Bi_2Te_3 at T = 78°K [44, 45] indicated that the energy surfaces of the valence and conduction bands of bismuth telluride are ellipsoids centered on the symmetry planes (Fig. 6). A similar result was obtained theoretically [46]. More recently it has been shown that the six-ellipsoid structure is in good agreement with galvanomagnetic measurements on p- and n-type Bi_2Te_3 in the temperature range 4-300°K [47, 48]. Unfortunately, the galvanomagnetic properties cannot by themselves be used to determine the absolute values of the effective masses; they give

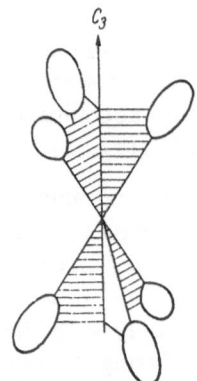

Fig. 6. Positions of the ellipsoids in Bi_2Te_3.

only the ratios $w_1 \simeq m_2/m_1$, $w_2 \simeq m_2/m_3$, as well as the angle of rotation θ of the ellipsoids with respect to the coordinate axes. Moreover the solution is not unique and the true solution cannot be identified since both values satisfy the results. The table lists the anisotropy parameters of the conduction and valence band ellipsoids at T = 78°K [48]. The table shows that the anisotropy parameters of the energy spectra of n- and p-type Bi_2Te_3 do not differ greatly.

The results of measurements of the piezoresistance effect in the cleavage plane of single-crystal samples of p- and n-type Bi_2Te_3 at 78-300°K confirm in general the validity of the six-ellipsoid model [49, 50]. However, when the piezoresistance is measured parallel to the C_3 axis, the effect is not equal to zero, which disagrees with theoretical predictions [51] based on the six-ellipsoid model. The reason for this is not yet clear.

Bismuth Selenide. The band structure of bismuth selenide is practically unknown. The only data which give some information on it are those on the piezoresistance of n-type Bi_2Se_3 with a high electron density [50]. The piezoresistance effect is very small, which seems to indicate that the conduction band of Bi_2Se_3 consists of ellipsoids of revolution with their centers on the C_3 axis. More reliable information may come from measurements on samples with lower carrier densities.

Conclusions

In conclusion it should be noted that as yet very little is known about the form of the carrier energy spectra of the majority of thermoelectric materials. The reason for this is mainly the unavailability of single crystals with uniform composition. Because of this the problem of preparing perfect single crystals with various carrier densities becomes particularly urgent.

LITERATURE CITED

1. B. Lax, Uspekhi Fiz. Nauk 70:1/2, 1960.
2. G. Dresselhaus et al., In: Problems of Semiconductor Physics [Russian translation], IL, Moscow, 1957, p. 599.
3. C. S. Smith, Phys. Rev. 94: 42, 1954.
4. F. J. Morin et al., Phys. Rev. 105:525 1957.
5. Yu. V. Ilisavskii, Fiz. Tverd. Tela 4: 918, 1962.
6. C. Herring and E. Vogt, In: Problems of Semiconductor Physics [Russian translation], IL, Moscow, 1957, p. 567.
7. G. E. Pikus and G. L. Bir, Fiz. Tverd. Tela 1: 1642, 1959.
8. W. Paul, J. Appl. Phys., Suppl., 32: 2082, 1961.
9. A. A. Averkin, B. Ya. Moizhes, and I. A. Smirnov, Fiz. Tverd. Tela 3: 1859, 1961.
10. E. F. Gross and A. A. Kaplyanskii, Fiz. Tverd. Tela 2: 2986, 1960.
11. E. F. Gross, A. A. Kaplyanskii, and V. G. Agekyan, Fiz. Tverd. Tela 4: 1009, 1962.
12. R. W. Keyes and M. Pollak, Phys. Rev. 118: 1001, 1960.
13. R. J. Stiles et al., J. Appl. Phys., Suppl., 32: 2174, 1961.
14. F. Herman, Phys. Rev. 95: 847, 1954.
15. H. Y. Fan, Uspekhi Fiz. Nauk 733, 1958.
16. R. Bowers, Phys. Rev. 108: 683, 1957.
17. H. Brooks, Probl. Sovr. Fiz. (8): 74, 1957.
18. E. M. Conwell, Proc. Inst. Radio Engrs. 46: 1281, 1958.
19. R. A. Laff and H. Y. Fan, Phys. Rev. 112:317, 1958.
20. A. Levitas et al., Phys. Rev. 95: 846, 1954.
21. A. Levitas, Phys. Rev. 99: 1810, 1955.
22. R. Braunstein et al., Phys. Rev. 109: 695, 1958.
23. G. Dresselhaus et al., Phys. Rev. 100: 1218, 1955.
24. M. Glicksman, Phys. Rev. 100: 1146, 1955.
25. M. Glicksman, Phys. Rev. 100: 1258, 1955.

26. M. Glicksman and S. M. Christian, Phys. Rev. 104: 1278, 1956.
27. F. Herman, Phys. Rev. 95: 847, 1954.
28. T. E. Slykhouse and H. G. Drickamer, J. Phys. Chem. Solids 7: 210, 1958.
29. I. N. Marshall and W. Paul, Phys. Rev. Letters 7: 52, 1961.
30. K. Shogenji and S. Uchiyama, J. Phys. Soc. Japan 12: 1164, 1957.
31. R. S. Allgaier, Phys. Rev. 112: 828, 1958.
32. K. Shogenji, J. Phys. Soc. Japan 14: 1360, 1959.
33. R. S. Allgaier, Phys. Rev. 119: 554, 1960.
34. R. S. Allgaier, Proc. Intern. Conf. on Semicond. Phys., Prague, 1037, 1960.
35. L. E. Hollander and T. J. Diesel, J. Appl. Phys. 31: 692, 1960.
36. R. J. Burke et al., Bull. Am. Phys. Soc., Ser. II, 6: 136, 1961.
37. P. J. Stiles et al., Phys. Rev. Letters 6: 667, 1961.
38. K. F. Cuff et al., J. Appl. Phys., Suppl., 32: 2179, 1961.
39. G. L. Bir and G. E. Pikus, Fiz. Tverd. Tela 3: 3050, 1961.
40. E. D. Devyatkova and I. A. Smirnov, Fiz. Tverd. Tela 3: 2310, 1961.
41. G. L. Bir and G. E. Pikus, Fiz. Tverd. Tela 4: 2243, 1962.
42. Yu. V. Ilisavskii, Fiz. Tverd. Tela 4: 1975, 1962.
43. J. R. Drabble and R. Wolfe, Proc. Phys. Soc. (London) 69: 1101, 1956.
44. J. R. Drabble et al., Proc. Phys. Soc. (London) 71: 430, 1958.
45. J. R. Drabble, Proc. Phys. Soc. (London) 72: 380, 1958.
46. E. K. Kudinov, Fiz. Tverd. Tela 3: 2, 1961.
47. B. A. Efimova et al., Fiz. Tverd. Tela 3: 2746, 1961.
48. B. A. Efimova et al., Fiz. Tverd. Tela 4: 302, 1962.
49. Yu. V. Ilisavskii, Fiz. Tverd. Tela 3: 1898, 1961.
50. Yu. V. Ilisavskii, Fiz. Tverd. Tela 4: 818, 1962.
51. M. I. Klinger, Fiz. Tverd. Tela 2: 1353, 1960.

METHODS OF MEASURING THE THERMAL CONDUCTIVITY
OF SEMICONDUCTORS AT HIGH TEMPERATURES

A. V. Petrov

The main difficulties in measuring the thermal conductivity by the steady-state method at high temperatures are related to the determination of the effective heat flux through the sample, measurement of the temperature gradient in the sample, and the chemical activity of the substances tested.

The methods of measuring the thermal conductivity at high temperatures differ mainly in the means of determining the effective heat flux producing the measured temperature drop. It is essential to ensure conditions such that the temperature gradient in the sample is constant, in other words, to ensure that losses from the side surface should be a minimum. Such losses are very difficult to allow for and can give rise to considerable error.

The simplest method is an absolute one, in which the radiative heat losses are reduced by means of radiation screens and the remaining losses are allowed for. The method is illustrated schematically in Fig. 1. The sample 1 together with a heater 2 is fixed to the base 3 of the apparatus; the base acts as a heat sink. A screen 4 is above the sample and heater; the temperature of the whole surface of the screen is constant. The radiative heat losses are calculated for this temperature.

The optimum positions of the radiation screens are such that the gap between the radiating surfaces and the screens is small compared with the geometrical dimensions of the sample and the heater. The screens should be made from good reflectors. However, there is a radiation flux parallel to the sample, between the side surface of the sample, across which there is a temperature gradient, and the screens. A calculation of this flux is given in the appendix to [1]. An estimate of this flux for an actual case shows that at about 1000°K it may amount to 30-50% of the total heat flux through the sample.

If one bears in mind that 1 cm^2 of the surface of a body with an emissivity of 0.5 of the emissivity of an absolute black body loses almost $6 \cdot 10^{-3}$ cal/sec for a temperature drop of 1° (at T = 1000°K), it becomes clear that great care is needed in using this method at high temperatures. Further complication arises from the lack of data on the integral emissivities for most materials.

A radiation screen was used by Abeles [2] in a study of the thermal conductivity of Ge to 800°. The use of this method can be justified for Ge because of its high thermal conductivity.

The radial heat flow method is also an absolute one. It is in principle the most reliable method among all the steady-state methods for high temperatures because of the absence of heat losses. The effective heat flux can then be determined most simply. This is done as follows (Fig. 2). A heater is placed along the axis of a cylindrical sample; the heater produces a temperature gradient in the radial direction. The gradient is measured with two thermocouples placed along a radius. The thermal conductivity is calculated from the following formula:

$$\kappa = \frac{q \cdot \ln\left(\frac{r_2}{r_1}\right)}{2\pi \left(T_1 - T_2\right)},$$

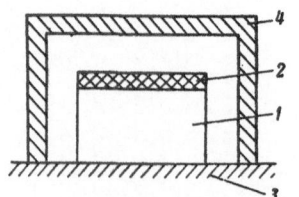

Fig. 1. Apparatus with a radiation screen.

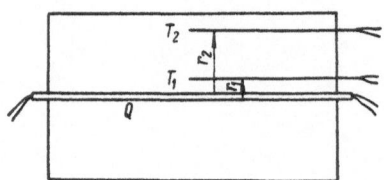

Fig. 2. Apparatus using radial heat flow. Q is the heater and T_1 and T_2 are thermocouples.

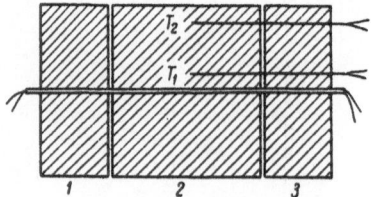

Fig. 3. Apparatus using radial heat flow and a composite sample.

where q is the electrical power per unit length of the heater; r is the distance from the center of the sample to a thermocouple; T is the temperature at a thermocouple.

To reduce distortions of the temperature field at the ends of the sample to below 2-3%, the ratio of the sample length L to its diameter d should be $L/d \approx 4$. This rigorous requirement may be met by using sectioned samples (Fig. 3). This produces a considerable thermal resistance in the longitudinal direction and improves the temperature distribution in the central part 2. The guard cylinders (1, 3) may be made of another material with a thermal conductivity and temperature dependence similar to that of the sample. The required sample dimensions still remain fairly large, since two thermocouples have to be placed along the radius and they should be separated by 5-6 mm. From this it follows that it is difficult to make the sample less than 2 cm in diameter.

The main error in the absolute value of the thermal conductivity is due to the inaccuracy of the measurement of the distance from the sample axis to the thermocouples. This may easily reach high values of 10-20% (as found in [3] in measurements of the thermal conductivity of Ge). Therefore the absolute value of the thermal conductivity should be checked by a different method.

When only two thermocouples are used, errors may easily arise in the temperature dependence of κ for two reasons:

1. A change in the contact between the heater and the sample leads to a disturbance of the temperature field symmetry. This alters the values of the calculated heat flux.

2. Slight changes in the position of the inner thermocouple may affect considerably the measured value of the temperature. This is related to the considerable temperature gradient over a distance equal to the dimensions of the thermocouple junction.

Thus the problem of mounting the thermocouples and the heater meets with practical difficulties. The sample, together with the heater and thermocouples, is the main part of the measuring system and should be re-assembled before each new measurement. This important disadvantage of the radial method is avoided by using a sample cut into two halves along the cylinder axis. Assembly of the heater and the thermocouples is then simplified considerably, and they may be used for measurements on several samples.

In measurements by the radial method the thermocouples are placed at right angles to the temperature gradient to ensure the correctness of their readings. This, together with the possibility of establishing conditions such that the radiation from the heater is small, is one of the advantages of the radial method. This is essential in measurements of the thermal conductivity of substances which are transparent in the infrared region of the spectrum. At a high temperature the electromagnetic radiation flux through samples of such substances may be comparable or greater than the energy flux due to the thermal conductivity of the lattice. The small surface area of the heater makes it possible to reduce this additional mechanism of heat transfer.

At temperatures exceeding 1000-1200° the radial method, in one or other of its modifications, is the one mainly used. Temperatures are measured with high-temperature thermocouples or with optical pyrometers.

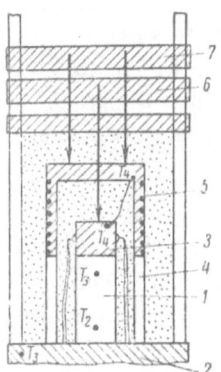

Fig. 4. Main part of the apparatus for measuring the thermal conductivity by an absolute method at high temperatures.

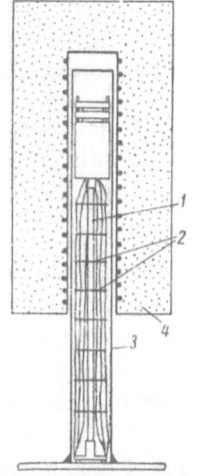

Fig. 5. General appearance of the apparatus .

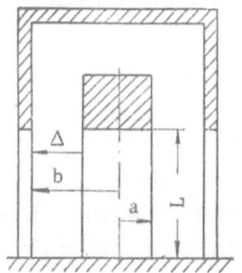

Fig. 6. Schematic representation of the apparatus.

The comparison method gives the required value of the thermal conductivity by comparison with a known value. This is done by placing it series two samples with known and unknown thermal conductivity and passing a heat flux q through them. Then, from the equality of heat flow in the sample and in the standard, we obtain

$$\kappa = \kappa_0 \frac{S_0}{S} \frac{l}{l_0} \frac{\Delta T_0}{\Delta T},$$

where κ is the thermal conductivity; S is the transverse cross-sectional area; and ΔT is the temperature drop over a length l. The quantities κ_0, S_0, l_0, ΔT_0 refer to the standard.

Essentially, one equalizes the temperature gradients in the sample and in the standard. It is then assumed that the heat flux through the sample and through the standard is the same. In practice this is difficult to achieve even with a system of screens. Therefore it is always necessary to use two standards, placing the sample between them. The heat flux through the sample is taken to be the average of the fluxes through the first and second standards.

The heat flux through the sample is not necessarily equal to the heat flux through the standards even if, by means of screens, the temperature gradients in the first and second standards are made equal. If the thermal conductivities of the standards κ_0 and of the sample κ_S are not equal, heat will tend to bypass the region with low conductivity by flowing along the thermal insulation. Thus the measured value of the thermal conductivity of the sample, κ_S', will lie between the values κ_S and κ_0. The error can be estimated from [4]

$$\Delta \kappa \equiv \kappa_S - \kappa_S' \approx m \kappa_m \left(\frac{\kappa_S'}{\kappa_0} - 1 \right),$$

where κ_m is the thermal conductivity of the insulation, and m is a geometrical factor equal to 2.

The main difficulty in the use of the comparison method at high temperatures is the absence of standards. The thermal conductivity of some types of steel and of several metals which have considerable conductivity is well known, but no measurements by absolute methods have been made on substances of low thermal conductivity in the range from room temperature to 1000°. An attempt to use a fused quartz standard [1] was unsuccessful because later measurements showed that due to the transparency of quartz at wavelengths shorter than 3 μ its thermal conductivity at temperatures higher than 300-400° depends on the conditions under which the measurements are carried out.

An absolute method for measuring the thermal conductivity between room temperature and 1000° of substances used in thermoelectric devices has been developed at the Institute of Semiconductors of the USSR Academy of Sciences.

The principle of the method is clear from Fig. 4. The sample 1 with a heater 3 is placed on the base 2 of the apparatus. A system of screens consists of a ring 4, equal in height to the sample, and a screen 5 which has another heater. During measurements the main heater establishes a small temperature drop across the sample (5-10°) which is measured with two platinum–platinum/rhodium thermocouples. While measurements are being made, the screen 5 is automatically kept at the temperature of the heater 3. This is checked with a differential thermocouple, one junction of which is in the heater 3 and the other in the screen 5. In this way, upward and side heat losses from the heater 3 are avoided. The temperature drops in the sample and in the ring are then equal, and there is an equality of temperatures in the same cross sections. This avoids heat losses from the sides of the sample.

All the internal space between the screen and the sample is filled with a thermal insulating powder, which prevents heat loss by radiation. Then, however, there is a heat flow parallel to the sample along the filter, part of which, Δq, is taken from the heater 3. The thermal conductivity is calculated by means of the standard formula for the steady-state case:

$$\kappa = \frac{(W - \Delta q)\, l}{S \cdot \Delta T},$$

where $W = UI$ is the electrical power consumed by the heater 3; l is the distance between the thermocouples; S is the cross-sectional area of the sample; ΔT is the temperature drop in the length l.

The heater 3 and the thick screen 5 are made of nickel in order to ensure good temperature equalization. The same purpose is served by the thermal insulation outside the screen.

All the contact surfaces must be carefully ground to reduce the temperature drop and to make it the same at the ends of the sample and the ring. To improve the thermal contact and to avoid its variation with temperature, the heater 3 is pressed against the sample, and the screen against the ring, by means of weights 6 and 7.

The thermocouples are attached to the sample as follows. Holes are drilled right through the sample, their diameters being such that platinum rods of 0.5 mm diameter fit them closely. Thermocouples are attached to the projecting ends of the rods. Each platinum rod assumes the temperature of the sample cross section in which it is located and its temperature is measured with the thermocouple. To ensure that the measured temperature is correct, each thermocouple is wound once or twice around the sample at the level of the cross section in which the rod is located. The distance l, which enters the formula for the calculation of the thermal conductivity, is measured between the centers of the rods.

The thermoelectric power α is determined, simultaneously with the thermal conductivity, from the emf E_α between the platinum ends of the thermocouple. Hence the absolute thermoelectric power of the substance is found from

$$\alpha_{abs} = \frac{E_\alpha}{\Delta T} \pm \alpha_{Pt},$$

where the sign depends on the type of conduction in the test substance.

The electrical conductivity can be measured together with the thermoelectric power. For this purpose thin platinum plates are placed below and above the sample and these are used as electrodes. They also protect the sample from possible chemical reaction between the material of the sample and the heater or the base of the apparatus.

The apparatus is fixed to a porcelain tube 1 (Fig. 5). A series of screens 2 is located on the tube and these are also used as supports for all the leads from the heaters and thermocouples to the base of the apparatus. The apparatus is sheathed outside with a quartz cover 3. The space within the cover is first evacuated to fore-vacuum pressure and then filled with argon; measurements are carried out in argon at atmospheric pressure. It is better to carry out experiments in argon than in vacuum for two reasons: the temperature drops at the contacts are considerably reduced, which improves the measurement conditions; evaporation of the sample material at high temperatures is also reduced. A further heater 4 is placed around the quartz cover; this is used to produce the required temperature between room temperature and 1000°.

We shall now consider the possible systematic and random errors of this method.

1. There is a heat flow parallel to the sample through the filter. This flow exists under all conditions of measurement. The quantity Δq represents that part of the total heat which flows through the filler from the heater and the sample. The closer the screen and the ring to the sample, the smaller Δq, because of the reduction of the effective cross section of the filler. The value of Δq is given by the following expression (obtained by B. Ya. Moizhes):

$$\Delta q \simeq \pi \cdot a \cdot \Delta \frac{\left(\frac{b}{a}\right)^{1/2} + 2}{3} \kappa_f \cdot \frac{\Delta T}{L},$$

where ΔT is the temperature drop over the whole length L of the sample, and κ_f is the thermal conductivity of the filler. The meaning of the other symbols is explained in Fig. 6.

The heat Δq flows from those regions of the sample and the heater which lie $\sim \Delta/2$ above and below the top of the sample. Therefore the upper thermocouple in the sample should be placed at a distance $\sim \Delta/2$ from its top. Then the heat $W - \Delta q$ does indeed flow through both cross sections of the sample in which the thermocouples are located. Hence it follows that the relationship between L and Δ should be $L \geq 2\Delta$ if the thermocouples are placed at a distance $L/4$ from the sample top.

2. If the temperatures of the screen and the heater differ by δT, heat will flow from the sample and the heater to the screen and the ring (or in the opposite direction). The value of this flow can be estimated [4]:

$$q_1 \sim \kappa_f \frac{2aL}{\Delta} \delta T \quad \text{(from sample to ring)},$$

$$q_2 \sim \kappa_f \frac{4a^2}{\Delta} \delta T \quad \text{(from heater to screen)}.$$

This flow may be made sufficiently small compared with the main heat flow through the sample if the differential thermocouple readings are checked periodically and the temperature drop across the sample is not too small.

3. An error arises in the case when the temperature drops at the ends of the sample and the ring are not the same. This is more difficult to estimate. The change in the heat flow through the sample is then approximately equal to

$$q_3 \sim \kappa_f \frac{2a}{\Delta} \kappa_s \ W_k \cdot \Delta T,$$

where W_k is the difference between the heat resistances of the sample and ring ends. Usually W_k does not exceed 15-20 $cal^{-1} \cdot cm \cdot sec \cdot deg$. In order to reduce this error, one should make the thermal contacts identical at the sample and ring ends. Moreover, the thermal conductivities of the ring and sample material should be quite similar. Since the errors in heat flow are greatest in the case of samples with low thermal conductivity, it is an advantage to make the ring of a material of low thermal conductivity. For this purpose fused quartz and porcelain are useful.

An estimate of the systematic and random errors of the method and the experience acquired in the use of the apparatus show that the errors in the measured absolute value of the thermal conductivity do not exceed 5-7%, and those in its temperature dependence are not more than 3-4%.

Some aspects of the thermocouples should be mentioned. The platinum –platinum/rhodium thermocouples have high stability. Thermocouples made from the same batch of platinum and platinum/rhodium differ very little from one another. However, since relatively small temperature drops are measured, this difference in calibration may give rise to a considerable error in ΔT. To avoid this, one or two measurements at high temperatures should be carried out using different heat flows through the sample. It is then possible to determine and allow for the difference in the thermocouple calibrations quite easily. Since this difference increases linearly with temperature, it may be determined and allowed for in the whole range of measurements.

LITERATURE CITED

1. E. D. Devyatkova, A. V. Petrov, I. A. Smirnov, and B. Ya. Moizhes, Fiz. Tverd. Tela 2: 738, 1960.
2. B. Abeles, J. Phys. Chem. Solids 8: 340, 1959.
3. G. Clack, Phys. Rev. 120: 782, 1960.
4. R. Heikes and W. Ure, Thermoelectricity, Science and Engineering, New York-London, 1961.

METHOD FOR THE RAPID DETERMINATION
OF THE TEMPERATURE DEPENDENCE
OF THE ELECTRICAL CONDUCTIVITY
OF SEMICONDUCTORS

V. B. Antonov and R. Kh. Nani

Investigation of the temperature dependence of the electrical conductivity is important in the study of the semiconducting materials. Measurement of the electrical conductivity is a well-developed subject and very accurate results can be obtained. However, these measurements, carried out usually by the compensation method, involve the necessity of stabilizing the temperature at the moment of measurement, which is both difficult and takes considerable time, especially at low temperatures. In practice, however, it is frequently necessary, without bothering too much about the high accuracy of the numerical values, to determine the temperature dependence of the electrical conductivity in a wide range of temperatures. Then the compensation method is not very useful.

Here we propose a method for rapidly determining the temperature dependence of the electrical conductivity of semiconductors in a wide range of temperatures. The method is based on recording the results of the measurements by means of recording instruments. Continuous recording avoids the necessity of temperature stabilization and speeds up the measurements.

In the probe method, the electrical conductivity σ at a given temperature is determined from the formula

$$\sigma = \frac{Il}{VS},$$

where I is the current flowing through the sample; V is the voltage drop between the measuring probes; l is the distance between the probes; S is the cross-sectional area of the sample.

The voltage drop across the probes is usually measured for both directions of the current through the sample in order to avoid the influence of stray thermo-emf's.

To make the measuring process automatic it is necessary to record simultaneously four quantities: temperature, current, and voltage drop for both directions of the current.

In our apparatus (Fig. 1) this is done as follows. The voltage drop across the probes for both directions of the current and the thermocouple readings are fed alternately, by means of an automatic commutator, to the input of a multiscale recording millivolt-microammeter of N-373/2 type, operating with an optical compensation circuit and having a zero in the middle of its scale. In this way marks are obtained on the recorder chart which, when joined, give the curves of variation of the voltage drop across the probes for both directions of the current (to the right and left from zero), and a curve showing the temperature variation. Two complete cycles of measurements (V_1, V_2, t) can be carried out in one minute.

A sufficiently large resistance is connected in series with the sample so that the current remains practically constant. To measure and check the constancy of the current, its value is

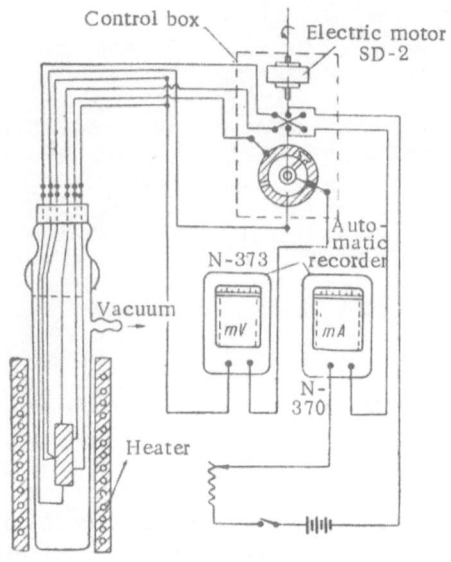

Fig. 1. Diagram of the apparatus.

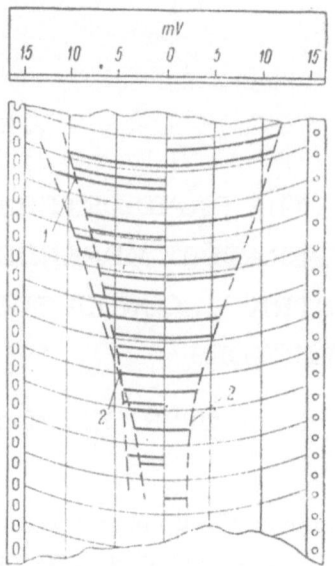

Fig. 2. Specimen record of the temperature dependence of the electrical conductivity. 1) Temperature variation curve; 2) voltage drop across the probes.

recorded in synchronism with the voltage drops and temperature, using a recording milliammeter of N-370 type.

The commutator is in the form of a switch rotated by an electric motor SD-2 at a rate of 2 rpm; it ensures successive connection of the thermocouple and probe circuits, and it reverses the current direction through the sample.

A typical record is shown in Fig. 2, where for clarity the record is compressed along its length. The record is read with a scale ruler and the value of σ is calculated for each temperature in the usual way.

The sample, fixed in a holder, is placed in a quartz tube evacuated to 10^{-4} mm Hg. Depending on the range of temperatures required, the tube is placed in a Dewar with liquid nitrogen or in a resistance furnace.

Measurements carried out in this way give results in good agreement with those obtained by the usual potentiometric circuit. The accuracy of the method depends on the accuracy of the recording instrument (class 1.5) and possible inaccuracies in interpreting the record.

The method makes it possible to carry out measurements in the range 100-700°K within a few hours.

ELIMINATION OF SOME ERRORS CONNECTED
WITH THE IRREPRODUCIBILITY OF THERMOCOUPLES

S. V. Airapetyants

Measurements of small temperature differences at high average temperatures require a high uniformity of the material used for making the measuring thermocouples.

Let the thermoelectric powers α_I and α_{II} of thermocouples I and II (cf. the figure) differ by a small quantity $\delta\alpha = \alpha_I - \alpha_{II}$. When these thermocouples are used to measure a small temperature difference $\Delta T = T_2 - T_1$, the relative error $\delta T/\Delta T$ in the determination of ΔT is

$$\frac{\delta T}{\Delta T} = \frac{\overline{T}\delta\alpha}{\Delta T \alpha}, \text{ where } \overline{T} = \frac{T_1 + T_2}{2}. \tag{1}$$

When $\Delta T = 1°$, $\overline{T} = 1000°$, and $\delta\alpha/\alpha_I = 0.05\%$, the error in the determination of ΔT amounts to 50%, according to Eq. (1).

The determination of small temperature differences ΔT is not, as a rule, done for its own sake but to calculate quantities which are derivatives with respect to temperature,

$$f(T) = \frac{dF(T)}{dT} \simeq \frac{\Delta F(T)}{\Delta T}. \tag{1a}$$

Examples of such quantities are thermoelectric power $\alpha = \Delta V/\Delta T$ and heat transfer coefficient $k = \Delta w/\Delta T$, where ΔV is the thermo-emf of the sample and Δw is the heat flow through the sample.

If $\Delta F(T)$ is determined for two values of ΔT at approximately the same temperature, then, denoting the values of ΔF and ΔT from the first and second measurements by the subscripts a and b, we have

$$f(T) = \frac{\Delta F(T)_a}{\Delta T_a} = \frac{\Delta F(T)_b}{\Delta T_b}. \tag{2}$$

From Eq. (2) we obtain

$$f(T) = \frac{dF(T)}{dT} = \frac{\Delta F(T)_b - \Delta F(T)_a}{\Delta T_b - \Delta T_a}. \tag{3}$$

Errors in the determination of small temperature differences by means of thermocouples with slightly different thermoelectric powers may be excluded by measuring ΔF at two different values of ΔT and at the same average temperature $\overline{T}$. Below we give the appropriate formulas for calculation. We shall use the notation α_1, α_2, α_3, α_4 for the algebraic values of the thermoelectric powers of the branches 1, 2, 3, 4 of the measuring thermocouples I and II (cf. the figure). Branches 1 and 3 are made of one material, and branches 2 and 4 from another; $\alpha_3 = \alpha_1 + \delta\alpha_1$, $\alpha_4 = \alpha_2 + \delta\alpha_2$, $\alpha_1 - \alpha_2 = \alpha_{1,2}$, the latter being the thermoelectric power of the thermocouple I; $\delta\alpha_1 - \delta\alpha_2 = \delta\alpha_{1,2}$ is the difference between the thermopowers of the thermocouples I and II; T_1 and T_2 are the temperatures of the hot junctions of the thermocouples I and II; $\Delta T = T_2 - T_1$; $V_{1,2}$, $V_{3,4}$, $V_{1,3}$, $V_{2,4}$ are the

Positions and notation of the thermocouples.

emf's between the cold ends of the corresponding branches of the thermocouples I and II; $\overline{T}$ = $(T_{1a} + T_{2a} + T_{1b} + T_{2b})/4$, T_0 is the temperature of the cold junctions of the thermocouples; α is the thermoelectric power of the sample on which ΔT is measured. We shall find $V_{1,2}$ and $V_{3,4}$ for the first measurement:

$$
\left.
\begin{aligned}
V_{1,2a} &= \int_{T_{0a}}^{T_{1a}} \alpha_{1,2}(T)\,dT, \\
V_{3,4a} &= \int_{T_{0a}}^{T_{2a}} \alpha_{3,4}(T)\,dT = \int_{T_{0a}}^{T_{1a}} \alpha_{1,2}(T)\,dT + \alpha_{1,2}\Delta T_a + \\
&\quad + \int_{T_{0a}}^{T_{2a}} \delta\alpha_{1,2}(T)\,dT.
\end{aligned}
\right\}
\tag{4}
$$

For the second measurement after a change of ΔT we obtain

$$
\left.
\begin{aligned}
V_{1,2b} &= \int_{T_{0b}}^{T_{1b}} \alpha_{1,2}(T)\,dT, \\
V_{3,4b} &= \int_{T_{0b}}^{T_{2b}} \alpha_{3,4}(T)\,dT = \int_{T_{0b}}^{T_{1b}} \alpha_{1,2}(T)\,dT + \alpha_{1,2}\Delta T_b + \\
&\quad + \int_{T_{0b}}^{T_{2b}} \delta\alpha_{1,2}(T)\,dT.
\end{aligned}
\right\}
\tag{5}
$$

From Eqs. (4) and (5) we obtain

$$
V_{1,2a} - V_{3,4a} - V_{1,2b} + V_{3,4b} = (\Delta T_b - \Delta T_a)\,\alpha_{1,2}(\overline{T}) + \\
+ [(T_{2b} - T_{2a}) - (T_{0b} - T_{0a})]\,\delta\alpha_{12}.
\tag{6}
$$

The difference given in brackets in the right-hand part of Eq. (6) is of the same order as $\Delta T_b - \Delta T_a$, but $\delta\alpha_{1,2}$ represents only hundredths of 1% of $\alpha_{1,2}$, and therefore the second term in the right-hand part of Eq. (6) can be neglected, giving us

$$
\Delta T_b - \Delta T_a = \frac{V_{1,2a} - V_{3,4a} - V_{1,2b} + V_{3,4b}}{\alpha_{1,2}(\overline{T})}.
\tag{7}
$$

If the hot junctions of the thermocouples are in electrical contact with the sample, we can measure $\Delta T_b - \Delta T_a$ by measuring the emf's $V_{1,3}$ and $V_{2,4}$ between the cold ends of the thermocouples I and II, made of the same material.

Let the current through the sample be I_a in the first measurement, and I_b in the second. Then

$$
V_{1,3a} = -\alpha_1(\overline{T}_a)\Delta T_a + \alpha(\overline{T}_a)\Delta T_a + I_a r(\overline{T}_a) - \int_{T_{0a}}^{T_{2a}} \delta\alpha_1(T)\,dT,
\tag{8}
$$

$$
V_{2,4a} = -\alpha_2(\overline{T}_a)\Delta T_a + \alpha(\overline{T}_a) + I_a r(\overline{T}_a) - \int_{T_{0a}}^{T_{2a}} \delta\alpha_2\,dT,
\tag{9}
$$

$$
V_{1,3b} = -\alpha_1(\overline{T}_b)\Delta T_b + \alpha(\overline{T}_b)\Delta T_b + I_b r(\overline{T}_b) - \int_{T_{0b}}^{T_{2b}} \delta\alpha_1\,dT,
\tag{10}
$$

$$V_{2,4b} = -\alpha_2(\bar{T}_b)\Delta T_b + \alpha(\bar{T}_b)\Delta T_b + I_b r(\bar{T}_b) - \int_{T_{0b}}^{T_{2b}} \delta\alpha_2 dT, \tag{11}$$

$$V_{1,3a} - V_{2,4a} - V_{1,3b} + V_{2,4b} = (\Delta T_b - \Delta T_a)\alpha_{1,2}(\bar{T}) +$$
$$+ (T_{2b} - T_{2a} + T_{0a} - T_{0b})\delta\alpha_{1,2}. \tag{12}$$

The right-hand part of Eq. (12) contains, with $\delta\alpha_{1,2}$ as the multiplier, a temperature difference of the order of $\Delta T_a(\Delta T_b)$. However, since $\delta\alpha_{1,2}$ represents only hundredths of 1 of $\alpha_{1,2}$, the term with $\delta\alpha_{1,2}$ can be neglected, and from Eq. (12) we obtain

$$\Delta T_b - \Delta T_a = \frac{1}{\alpha_{1,2}(\bar{T})}(V_{1,3a} - V_{2,4a} - V_{1,3b} + V_{2,4b}). \tag{13}$$

In Eqs. (7) and (13) errors in the determination of $\Delta T_b - \Delta T_a$ related to the irreproducibility of the measuring thermocouples are excluded. It is quite clear that, using Eq. (7), calculations can be carried out for the case when current flows through the sample. The same is true of Eq. (13), which can be used to determine ΔT when the current through the sample is different in the first and second measurements.

To calculate the heat transfer coefficient k from Eq. (3) we have

$$k = \frac{w_b - w_a}{\Delta T_b - \Delta T_a}, \tag{14}$$

where w_b and w_a are the given heat flows through the sample, one of which can be zero, and $\Delta T_b - \Delta T_a$ is determined from Eq. (7) or (13).

To calculate the thermoelectric power of the sample we use

$$\alpha(\bar{T}) = \frac{V_{1,3b} - V_{1,3a}}{\Delta T_b - \Delta T_a} + \alpha_1(\bar{T}) = \frac{V_{2,4b} - V_{2,4a}}{\Delta T_b - \Delta T_a} + \alpha_2(\bar{T}). \tag{15}$$

Here $\alpha_1(\bar{T})$ and $\alpha_2(\bar{T})$ are usually introduced to allow for the thermoelectric power of the leads between which the thermoelectric power of the sample is measured, and no current should flow through the sample in measurements of the thermoelectric power.

Since Eq. (14) has in its denominator the increment ΔT, the presence of a temperature gradient in the sample which is not produced by the heaters (w_b and w_a) but by the temperature gradient in the medium surrounding the sample should not affect the results. For the same reason the measurements of k and α are not affected by small changes of the average temperature of the sample between the measurements a and b.

LITERATURE CITED

1. A. I. Shtenbek and P. I. Baranskii, Zhur. Tekh. Fiz. 26: 1373, 1956.

MEASUREMENT OF THE THERMOELECTRIC PROPERTIES
OF SEMICONDUCTORS AT HIGH TEMPERATURES
BY THE HARMAN METHOD

S. V. Airapetyants

Harman proposed and others developed a method of measuring the thermoelectric figure of merit z of semiconducting materials [1, 2, 3]. The essence of the method lies in the following: A semiconducting sample is isolated thermally on thin current leads in vacuum (cf. the figure). On passing a direct current through the sample, a quantity of Peltier heat $Q_p = I\alpha T$ is emitted at one of the junctions of the semiconductor with metal electrodes 1 and 2, while at the other junction this heat is absorbed. Here I is the current, α is the thermoelectric power of the semiconductor, and T is the absolute temperature. The emission and absorption of Peltier heat at the contacts produces a temperature drop ΔT between the electrodes 1 and 2. Under steady-state conditions

$$Q_p = I\alpha T = k\Delta T, \tag{1}$$

where k is the heat transfer coefficient of the sample. Substituting $I = V_\rho \sigma s / l$, $k = \kappa s / l$, and $\alpha \Delta T = V_\alpha$ into Eq. (1) and multiplying by α, we obtain

$$zT = \frac{\alpha^2 \sigma}{\kappa} T = \frac{V_\alpha}{V_\rho}, \tag{2}$$

where σ is the electrical conductivity; κ is the thermal conductivity; s and l are, respectively, the cross section and length of the sample; V_ρ is the ohmic voltage drop across the sample; V_α is the total thermo-emf of the sample.

When a direct current passes through the sample, the voltage at the electrodes 1, 2 consists of the ohmic voltage drop V_ρ and thermo-emf V_α:

$$V_{1,2} = V_\rho + V_\alpha. \tag{3}$$

To separate $V_{1,2}$ into V_α and V_ρ it is necessary to measure the voltage at the electrodes 1, 2 immediately after switching off the current, when there is still a temperature difference across the sample. Then the measured voltage is V_α, and, knowing $V_{1,2}$, we can find V_ρ. We can also measure V_ρ using alternating current and then we can find V_α from Eq. (3).

The quantity z in Eq. (2) is a parameter which determines the efficiency of a thermoelectric generator and the coefficient of performance for thermoelectric cooling. Thus the Harman method permits direct measurements of the principal parameter of materials used for thermoelectric elements.

Moreover, knowing the temperature difference ΔT between the electrodes 1, 2, we can determine separately α, σ, and the thermal conductivity of the sample κ:

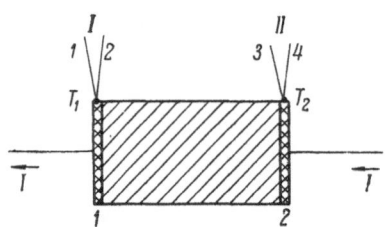

Positions of the sample, thermocouples, and current leads.

$$\alpha = \frac{V_\alpha}{\Delta T}, \quad \sigma = \frac{Il}{V_\rho s}, \quad \varkappa = \frac{I\alpha T}{\Delta T} \cdot \frac{l}{s}, \tag{4}$$

i.e., all the quantities which determine z.

The Harman method may also be used at high temperatures, but then the errors become greater for the following reasons.

1. Heat losses from the sample surface and from the electrodes increase due to radiation if the sample is in vacuum, or due to thermal conductivity if the sample is insulated by a filler in the form of a powder of low thermal conductivity.

2. At high temperatures it is more difficult to establish good electrical contact between the electrodes and the sample. When the contact resistances between the electrodes 1, 2 and the sample are different, an additional temperature difference appears across the sample due to the emission of Joule heat at the contacts. These errors may be reduced by reducing the current passing through the sample, but then the temperature difference across the sample ΔT will be small. Moreover, in the presence of temperature gradients in the heated enclosure, a temperature difference may appear across the sample even in the absence of a current.

3. If the material of the thermocouples I and II (cf. the figure) used to measure ΔT is not uniform, there will be large errors in the measurement of a small temperature difference ΔT across the sample. For example, if the thermocouples I and II give a 0.1-0.5° scatter of readings at 500°, then at $\Delta T = 1°$ the error in determining ΔT amounts to 10-50%.

Methods of avoiding these difficulties in using the Harman method at high temperatures are described below.

1. Experimental Determination of Heat Losses from the Sample. We shall consider the Harman method, allowing for heat losses from the sample. We shall use the following notation: T_0 is the temperature of the furnace in which the sample is placed; k_e is the total heat transfer between one of the electrodes and the walls of the furnace (this consists of the heat transfer along the thermocouples and current leads and the heat transfer by radiation and conduction through the filler and air); k_s is the total heat transfer between the side surface of the sample and the furnace wall by radiation and by transfer through the air or thermal insulation; $\overline{T} = (T_1 + T_2)/2$ is the average temperature of the electrodes 1 and 2. Harman et al. [1] showed that in the case of nonideal thermal insulation of the sample, Eqs. (2) and (4) should be replaced by the expressions

$$k = \frac{I\alpha T}{\Delta T} - \frac{k_e}{2} - \frac{k_s}{12}. \tag{5}$$

$$zT = \frac{V_\alpha}{V_\rho}\left(1 + \frac{k_e}{2} + \frac{k_s}{12}\right). \tag{6}$$

To determine k_e and k_s experimentally it is suggested that the electrodes 1 and 2 should have miniature heaters and special measurements of k_e and k_s should be carried out at different temperatures T_0. The value of k_e can be determined by placing the two electrodes in direct contact (without the sample), so that they are in good thermal contact, and switching on the electrode heaters. In the steady state

$$k_e = \frac{w_1}{2\,(\overline{T} - T_0)}, \tag{7}$$

where w_1 is the total power of the electrode heaters.

If this is done at different temperatures T_0, the temperature dependence of k_e can be determined.

To determine k_s, the sample should be placed between the electrodes and the electrode heaters switched on but without passing a current through the sample. In the steady state

$$k_s = \frac{w_2}{\overline{T} - T_0} - 2k_e. \tag{8}$$

Having determined k_s at various T_0, we can find the temperature dependence of k_s. If the same electrodes are used for measurements on different samples, the values of k_e should be the same for all samples.

2. __Formulas for Calculation.__ To eliminate the influence of the temperature drop due to the irreproducibility of the thermocouples, it is necessary to carry out measurements for both directions of the direct current. We shall denote all quantities measured with the current in the positive direction by the subscript a and use the subscript b for the negative direction of the current. Using the notation and formulas (7), (13), (14), (15) given in [4], we find

$$2\Delta T = \Delta T_b - \Delta T_a, \tag{9}$$

where ΔT is the quantity which occurs in Eq. (5) defining k:

$$\Delta T_b - \Delta T_a = \frac{1}{\alpha_{1,2}(\overline{T})} (V_{1,2a} - V_{3,4a} - V_{1,2b} + V_{3,4b}), \tag{10}$$

$$\Delta T_b - \Delta T_a = \frac{1}{\alpha_{1,2}(\overline{T})} (V_{1,3a} - V_{2,4a} - V_{1,3b} + V_{2,4b}). \tag{11}$$

We shall denote quantities measured when a current is flowing through the sample by the subscript I, and quantities measured immediately after the current is switched off, but while the temperature difference is still retained, by the subscript 0. We then find that

$$\alpha(\overline{T}) = \frac{V_{1,3b_0} - V_{1,3a_0}}{\Delta T_b - \Delta T_a} + \alpha_1(\overline{T}) = \frac{V_{2,4b_0} - V_{2,4a_0}}{\Delta T_b - \Delta T_a} + \alpha_2(\overline{T}), \tag{12}$$

$$2V_a = V_{1,3b_0} - V_{1,3a_0} + (\Delta T_b - \Delta T_a)\alpha_1(\overline{T}) = V_{2,4b_0} - V_{2,4a_0} + (\Delta T_b - \Delta T_a)\alpha_2(\overline{T}), \tag{13}$$

$$V_\rho = V_{1,3a_I} - V_{1,3a_0} = -V_{1,3b_I} + V_{1,3b_0} = V_{2,4a_I} - V_{2,4a_0} = -V_{2,4b_I} + V_{2,4b_0}. \tag{14}$$

Thus from Eqs. (9), (10), (11), (12) we find ΔT and α for the determination of k from Eq. (5); from Eqs. (13) and (14) we find V_α and V_ρ for the determination of zT from Eq. (6).

It should be noted that when these formulas are used small changes of the average sample temperature on reversal of the current do not affect the results.

In conclusion, it is necessary to point out that Eqs. (5) and (6) are valid if the total heat transfer from electrode 1 to the furnace walls surrounding the sample, k_{e1}, is equal to the heat transfer from the electrode 2, i.e., $k_{e1} = k_{e2}$. If this condition is not satisfied there may be errors in the determination of z and k given approximately by

$$\frac{T_0 - T_e}{\Delta T} \cdot \frac{k_{e1} - k_{e2}}{2k},$$

where $T_0 - T_e$ is the temperature difference between the furnace walls surrounding the sample and the electrodes; ΔT is the temperature drop across the sample due to the Peltier effect, given by Eq. (9); k is the heat transfer coefficient of the sample.

Consequently, in setting up the measuring apparatus the maximum symmetry with respect to the sample should be observed.

The following point should also be noted: In Eqs. (9)-(14) errors due to temperature change on current reversal are excluded, but if the sample temperature changes considerably during measurements for one of the current directions, then these errors are not excluded. Therefore, after switching on the current to the sample and establishing the steady state the whole measurement process should be carried out as quickly as possible.

LITERATURE CITED

1. T. C. Harman et al., J. Appl. Phys. 30: 1351, 1959.
2. E. K. Iordanishvili, Dissertation for Candidate's Degree, IPAN SSSR, 1961.
3. M. A. Kaganov, I. S. Lisker, and I. B. Mushkin, Fiz. Tverd. Tela 1: 988, 1959.
4. S. V. Airapetyants, present collection, p. 25.

REPRODUCIBLE THERMAL PRESSURE CONTACTS

S. V. Airapetyants and M. N. Ryabinin

The main errors in measurements of thermal conductivity, especially at high temperatures, are due to heat losses from the surface of the heat source and the sample. To reduce
these losses one can use short samples (for example, 0.1 cm long and 1 cm² cross section).

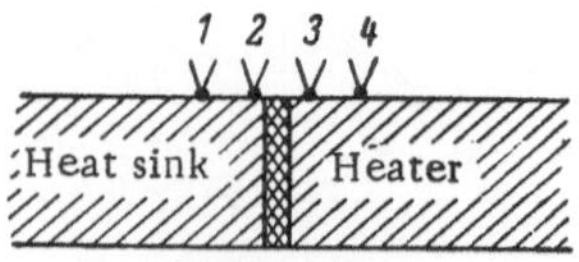

Fig. 1. Positions of the thermocouples.

However, it is then necessary to ensure that the thermal contacts
are sufficiently good. The thermal resistance of the contacts
should have a known and reliably reproducible value or it should
be small compared with the thermal resistance of the sample.

To obtain reproducible thermal contacts we propose here the
use of graphite powder. Narrow gaps (0.2-0.3 mm) are left between the ends of the sample, heater, and heat sink and these gaps
are filled with graphite powder.

To check the magnitude and reproducibility of the thermal
resistance of these contacts, special experiments were carried out.
The powder was pressed into the gap between the ends of two brass rods (heater and heat sink
in Fig. 1), and thermocouples 1-4 were used to measure the thermal resistance in the gap.
Measurements were carried out at room temperature and with pressures of 6 and 25 kg/cm²
applied to the powder. The table lists the values of the thermal conductivity of a contact layer
0.3 mm thick (average values of 10-20 measurements) and the root-mean-square deviations.
The last column of the table gives the ratios of the thermal resistances of the contacts and of
a sample for which $\kappa = 5 \cdot 10^{-3} \text{cal} \cdot \text{cm}^{-1} \cdot \text{sec}^{-1} \cdot \text{deg}^{-1}$ and which is 1 mm long.

The table shows that the uncontrollable part of the contact thermal resistance represents
respectively 0.7, 0.3, and 0.7% relative to the sample. Thus the error due to the scatter of
the contact thermal resistance values does not exceed 1%.

Moreover, if necessary, the thermal resistance of a contact can be determined more exactly in each case by measuring its electrical resistance. Figure 2 gives the relationship between the electrical and thermal resistances for two groups of measurements corresponding
to the second and third rows of the table. Figure 2 shows that the electrical resistance of a

Thermal Properties of Contacts

Contact-layer material	Pressure, kg/cm²	Thermal conductiv-ity,cal·cm⁻¹·sec⁻¹·deg⁻¹	r.m.s. deviation, %	Correction for contact layer, %
Ungraded graphite powder	6	0.0095	2.1	31
Ditto	25	0.019	2.1	16
Graphite grain dimen-sions 0.1-0.3 mm . .	6	0.0085	1.9	35

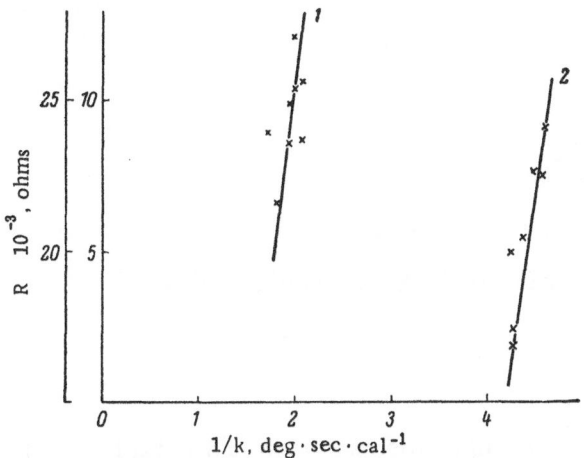

Fig. 2. Relationship between the thermal resistance and the electrical resistance of a contact layer 0.03 cm thick. 1) Powder grain dimensions 0.01-0.03 cm, pressure 6 kg/cm²; 2) ungraded powder, pressure 25 kg/cm².

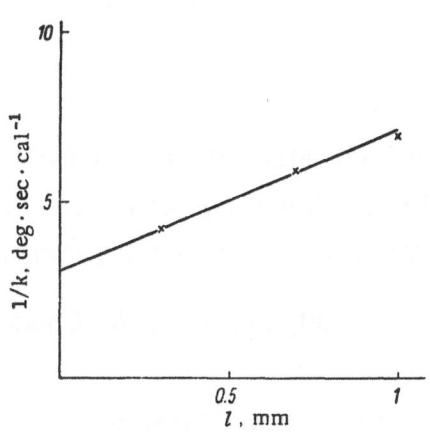

Fig. 3. Dependence of the thermal resistance of the contact layer on its thickness.

contact layer is proportional to its thermal resistance with an accuracy sufficient for our purpose. The scatter of points away from a straight line decreases with increase of the pressure and particularly after grading the powder. Thus, for the first group of measurements (not given in Fig. 2) the maximum deviation of an experimental point from a straight line amounts to 11% and the root-mean-square deviation is 2%; for the second group these values are 5.3 and 1.4%, while for the third group they are 2.4 and 0.4%. The slope of the straight line increases with pressure but falls rapidly on increase of the contact layer thickness.

Figure 3 shows the thermal resistance of a contact as a function of its thickness. A considerable part of the resistance of the layer is the intrinsic contact resistance. An important point here is that the layer thickness is comparable to the powder grain dimensions.

The advantage of this method is the rapidity of measurements, which is obtained at the expense of the requirements of steady temperature distribution. The latter is related to the short length of the sample; the ratio of the losses due to the specific heat of the sample to the main heat flow is proportional to the square of this length. If the sample is 1 mm long, $\kappa = 5 \cdot 10^{-3}$ cal $\cdot$ cm^{-1} $\cdot$ sec^{-1} $\cdot$ deg^{-1}, and its specific heat is 0.5 cal $\cdot$ cm^{-3} $\cdot$ deg^{-1}, then with a temperature drop of 10° across it, and a permissible error of 1%, the temperature can be varied at a rate of up to 360°/hr.

METHODS OF MEASURING THERMO-EMF'S
AND DESCRIPTION OF APPARATUS FOR MEASURING
THE INTEGRAL THERMOELECTRIC POWER OF SEMICONDUCTORS

A. I. Shelykh and V. Z. Chukanov

In investigations of thermoelectric phenomena in semiconductors and in their practical applications [1, 2, 3] it is necessary to know such quantities as the thermoelectric power α or the integral thermoelectric power E.

These parameters are related by the equations

$$\alpha = \frac{dE}{dT} \qquad (1)$$

and

$$E = \int \alpha \, dT, \qquad (2)$$

where T is the temperature; these parameters are determined by two main experimental methods [4, 5].

One of these methods, frequently called the differential method, is indicated schematically in Fig. 1, which gives the quantities being measured. The thermoelectric power in the variants a and b is given by the following formulas, respectively:

$$\alpha = \frac{\Delta E_1}{T_1 - T_2} \text{ and } \alpha = \frac{\Delta E_1}{\Delta E_2 - \Delta E_1} \cdot \alpha_{\text{thermocouple}}.$$

In experiments, a small temperature difference ΔT, of the order of several tens of degrees, is usually established between the ends of the sample. All the electrical quantities are measured potentiometrically. In the case of low-resistance samples, as well as in measurements of the thermocouple readings and the difference of these readings, both normal and differential galvanometers may be used.

The special construction features of the apparatus and the problems of the precision of measurements in the variant a are discussed in detail in [2, 6, 7]. The variant b was used by A. A. Rudnitskii in investigating thermoelectric effects in metals. In his apparatus the readings were recorded automatically on photographic paper by means of the three galvanometers of a Kurnakov pyrometer [8].

The temperature dependence of the integral thermoelectric power, $E = f(T)$, is determined in a different way (Fig. 2). In this case one of the sample contacts is kept at a constant temperature T_0 (for example, 0 or 20°C), and the other is heated to the required temperature T.

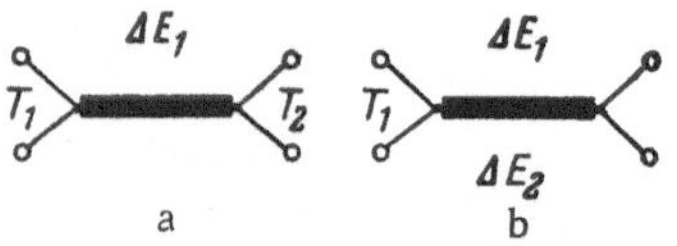

In the integral method, indicated schematically in Fig. 2a, two quantities are measured: E_1 and T. In the variant b (Fig. 2) the temperature is calculated from the difference of two variables: $E_1 - E_2 = E_{\text{thermocouple}} = f(T)$. The calibration curve of the

Fig. 1. Diagram illustrating the determination of the differential thermoelectric power.

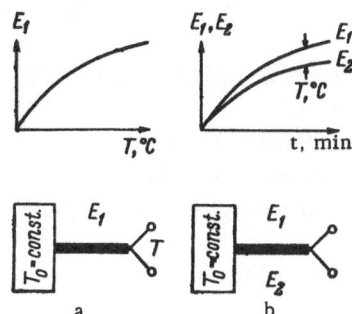

Fig. 2. Diagram illustrating the determination of the integral thermoelectric power.

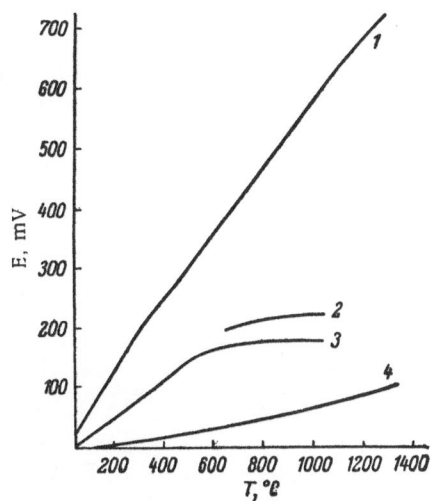

Fig. 3. Results of measurements of the integral thermoelectric power of some semiconductors in the liquid and solid state: 1) p-type NiO (platinum); 2) n-type Ge (Chromel); 3) n-type Ge (Alumel); 4) n-type SnO$_2$ (platinum).

thermocouple, i.e., the dependence $E_{thermocouple} = f(T)$ is assumed to be known.

Comparison of these methods from the point of view of the rapid automatic recording of the thermoelectric characteristics of semiconductors shows that it is preferable to use the integral thermoelectric power method. The main advantage of this method is that only two time-dependent quantities need be found: E and the corresponding hot-junction temperature T, because the temperature of the cold end remains constant at T_0. In many cases it may be an advantage that in measurements of the integral thermoelectric power at high temperatures only a part of the sample and not all of it is heated.

In order to check whether practical use may be made of the integral method in studies of the thermoelectric properties of semiconductors, we developed an apparatus [9] the working principle of which is illustrated in Fig. 2a.

The use of a thermostat at the cold end of the sample extends the possibilities of the integral method, which had been used as a rule only in studies of ductile materials from which long wires could be prepared.

We give below the main technical details of the apparatus intended for measuring the integral thermoelectric power of semiconductors in air. The value of the integral thermoelectric power E and the corresponding hot-end temperature T are recorded automatically on photographic paper. The time required to carry out measurements in the temperature range from 20 to 1400°C amounts to 15-30 min. The length of the measured samples is 30-40 mm. The apparatus is suitable for investigating semiconductor samples having resistances up to 10^6 ohms.

In measuring the thermoelectric power of low-resistance samples ($\simeq 10^2$ ohms) the results may be recorded directly with an automatic potentiometer EPP-09 or with a similar instrument.

Similar apparatus may be adapted for measurements in an inert-gas atmosphere or in vacuum. Figure 3 shows the results of measurements on samples of NiO and SnO$_2$ in air up to 1400°C and on Ge in argon at 10^{-2} mm Hg in a range of temperature which includes the melting point of germanium.

Because of the simplicity of this apparatus based on the principle of measurement of the integral thermoelectric power, and the possibility of automating the measuring process and recording the results, it may find applications not only in research work but in many problems connected with the manufacture and testing of thermoelectric materials. Among such problems are the determination of stability with time, stability under thermal cycling, detection of nonuniformities, aging, etc.

LITERATURE CITED

1. G. Busch and U. Winkler, Determination of the Characteristic Parameters of Semiconductors, in: Semiconductors in Science and Technology, Izd. Akad. Nauk, Moscow-Leningrad, 1958, Vol. II.
2. L. S. Stil'bans, Thermoelectric Phenomena, in: Semiconductors in Science and Technology, Izd. Akad. Nauk SSSR, Moscow-Leningrad, 1957, Vol. I.
3. A. F. Ioffe, Semiconductor Thermoelements, Izd. Akad. Nauk SSSR, 1960.
4. W. C. Dunlap, Introduction to Semiconductors [Russian translation], IL, 1959.
5. Metal Physics Encyclopedia, Ob'edinenie Nauch.-Tekh. Izd., Narodnyi Komis. Tyazh. Prom., SSSR, 1937, Vol. 1.
6. G. I. Skanavi and A. M. Kashtanova, Zhur. Tekh. Fiz. 26: 895, 1956.
7. B. I. Boltaks, Zhur. Tekh. Fiz. 20: 1039, 1950.
8. A. A. Rudnitskii, in: Thermoelectric Properties of Noble Metals and Their Alloys, Izd. Akad. Nauk SSSR, Moscow-Leningrad, 1956.
9. A. I. Shelykh and V. Z. Chukanov, Apparatus for Rapid Determination of the Integral Thermoelectric Power of Semiconductors in a Wide Range of Temperatures, Izd. PNTPO, No. 3, Subject 32, No. P-62-21/3, 1962.

DYNAMIC METHOD OF DETERMINING THE TEMPERATURE DEPENDENCE OF THE ELECTRICAL CONDUCTIVITY OF SEMICONDUCTORS

I. S. Lisker

At present the temperature dependence of the electrical conductivity of semiconducting materials is usually determined by the d-c probe compensation method, the temperature of the ambient medium being altered in steps. Such a method is inconvenient because a large amount of time is needed to record the $\sigma = f(T)$ curve and keep the sample temperature constant during measurements of the electrical conductivity. Obviously, this method involves measurement errors connected with recording, simultaneously with the ohmic voltage drop, the thermo-emf at the points of contact of the probes and the sample.

In view of this it would be of considerable interest to develop a method for determining the electrical conductivity at a fixed sample temperature, free of these disadvantages.

A nonstationary method for determining the electrical conductivity at a fixed temperature has been described earlier [1, 2]. The essence of the method is that, as a result of the inertia of the process of establishing a temperature field in the sample by the evolution of Peltier heat, it is possible to determine accurately the components of the total voltage drop at the probes and at the sample ends:

$$V_p = V_{p\rho} + V_{p\alpha},$$
$$V_e = V_{e\rho} + V_{e\alpha},$$

where $V_{p\rho}$, $V_{p\alpha}$ and $V_{e\rho}$, $V_{e\alpha}$ are the ohmic voltage drops and thermo-emf's at the probes (subscript p) and the sample ends (subscript e), respectively.

Figure 1 gives curves showing the variation of the voltage drop at the probes and across the sample. The rapidly rising parts of the curves represent the true ohmic voltage drop between the probes, $V_{p\rho}$, and across the sample, $V_{e\rho}$; the slowly rising parts of the voltage $V_{p\alpha}$ and $V_{e\alpha}$ are due to the thermo-emf. The voltage drops at the probes and at the sample ends are measured with a fast-acting automatic potentiometer (for example, EPP-09, single-point type, in which the time taken by the carriage to traverse the whole scale is 1 sec).

The method described makes it possible to determine σ of semiconducting materials, particularly thermoelectric materials, and to avoid error in the measurements due to the presence of parasitic thermo-emf's; thus the precision of measurements of the electrical conductivity is considerably increased.

The method suggested for measuring the temperature dependence of the electrical conductivity is based

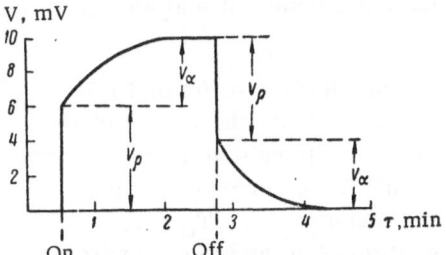

Fig. 1. Variation of the voltage drop across the sample on switching the current on and off.

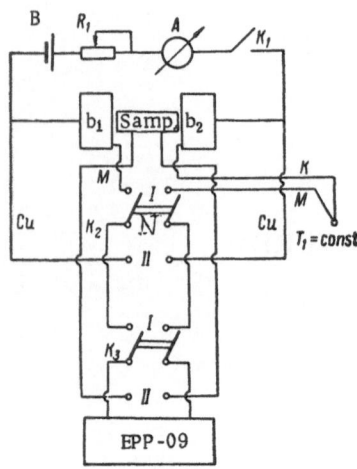

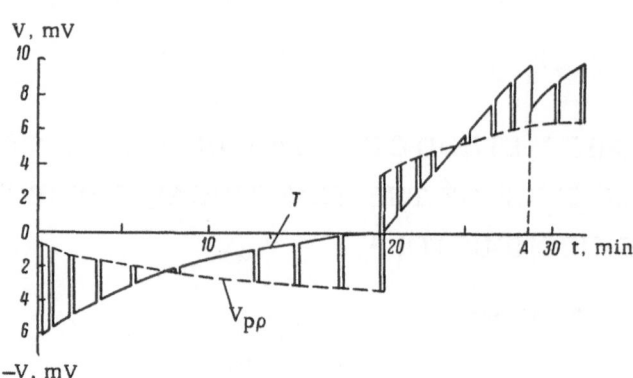

Fig. 2. Basic electrical circuit for measuring $\sigma = f(T)$ of low-resistance samples.

Fig. 3. Automatically produced record of the variation of the sample temperature and the voltage drop with time.

on the fact that simultaneously with the continuous automatic recording of the variation of the sample temperature with time (for example, during continuous heating from −195 to 500°C or higher), the ohmic voltage drop at the probes is recorded for various parts of the cooling or heating curve, i.e., at various sample temperatures.

The basic electrical circuit used in the measurements is shown in Fig. 2. The sample is placed between two massive copper blocks b_1 and b_2. The current circuit includes a battery B, a variable resistance R_1, an ammeter A, and a switch K_1. The sample temperature is measured with a differential thermocouple, one junction of which is short-circuited by the sample. It is obvious that this method of joining the thermocouple branches is possible only when the sample resistance is much smaller than the resistance of the thermocouple leads. The equality of temperatures along the sample is judged by the absence of a thermo-emf at its ends when the current is switched off (the switch K_2 is shifted to the position II, and the switch K_3 to the position I, while the switch K_1 is open).

To determine the voltage drop at the probes $V_{p\rho}$ for a given current i, the switches K_2 and K_3 are shifted to the neutral position and the position II, respectively; the current circuit is closed and then rapidly opened by the switch K_1. At the same time that the current circuit is broken the switches K_2 and K_3 are again shifted to the position for recording the sample temperature.

To measure the electrical conductivity continuously over a wide range of temperatures the sample with its holder is placed first in a cryostat and then in a heated chamber; the variation of the sample temperature with time is determined by means of an automatic recorder. This recorder determines the temperature difference between the sample and the medium in which the second thermocouple junction is located, the latter having a constant and known temperature.

Figure 3 shows a typical heating curve for a sample of thermoelectric alloy of Bi_2Te_3 − Bi_2Se_3 type. It is possible to measure σ in a range of temperatures outside the limits of the instrument scale (10 mV in EPP-09) by rapid variation of the fixed temperature T_C of the "cold" junction of the thermocouple (for example, in Fig. 3 up to the point A the junction temperature was $T_C = 22°C$ and then the thermocouple junction was placed in boiling water, $T_C = 100°C$). Single voltage-drop peaks in Fig. 3 correspond to instants of switching on and rapid switching off of the current through the sample. The curve joining the voltage drop peaks measured at the probes gives the temperature dependence of the electrical conductivity of the material.

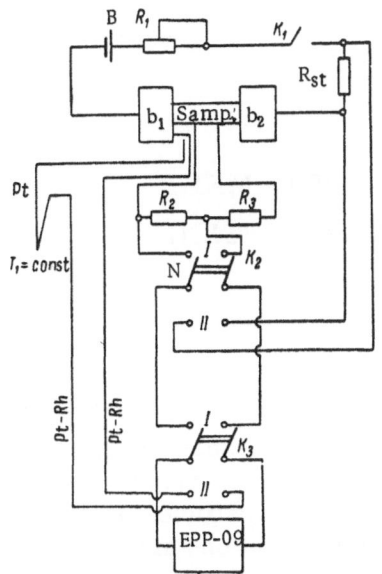

Fig. 4. Basic electrical circuit for measuring $\sigma = f(T)$ of high-resistance samples.

Knowing the values of the current i, the voltage drop at the probes $V_{p\rho}(T)$, the temperature T, the distance between the probes l_p, and the cross-sectional area of the sample S, we can easily determine the electrical conductivity of the sample at various temperatures.

In general, at each measurement of the voltage drop at the probes it is necessary to record simultaneously the current i flowing through the sample. This may be avoided in measurements of σ of thermoelectric materials by establishing some constant value of the current i before the experiment because its value is determined mainly by the ohmic resistance of other parts of the current circuit. Therefore, it is always possible to select a value of the rheostat resistance at which the temperature-induced changes of the sample resistance alter the current in the circuit so little that the changes are outside the limits of the ammeter sensitivity.

To determine the temperature dependence of the electrical conductivity of semiconducting materials with high resistivity (of the order of 10^3 ohm · cm) which may alter considerably during measurements (for example, materials with a large temperature coefficient of resistance), measurements should be made with the circuit shown in Fig. 4. This circuit is modified for measurements on high-resistance samples by including a single electronic potentiometer having a low input impedance (of the same type as in the circuit of Fig. 2). The current through the sample is determined by measuring the voltage drop across a standard resistance R_{st} and the voltage drop at the probes is found with a voltage divider R_2-R_3; in contrast to the circuit in Fig. 2, the sample temperature is determined by measuring the temperature of one of the current-supplying copper blocks.

The sample temperature variation is recorded by placing the switch K_2 in its neutral position and the switch K_3 in the position II (the switch K_1 is open). To measure the values of the current and the voltage drop at the probes, the switch K_3 is placed in the position I and the current circuit is closed by the switch K_1. Then the switch K_2 is first thrown over into the position I, and after several seconds to the position II. The circuit is then returned again to the position for recording the sample temperature.

The time required to determine the temperature dependence of the electrical conductivity depends on the rate of heating or cooling of the sample; it ranges between 10-20 min, depending on the temperature interval in which measurements are carried out.

The merits of the present method for measuring $\sigma = f(T)$ lie in the short duration of the experiment, high accuracy in determining σ (the error does not exceed ±1.5%), and the automatic recording of the measured quantities by means of a single, rapid-acting recording instrument.

LITERATURE CITED

1. M. A. Kaganov and I. S. Lisker, Zavodskaya Laboratoriya 10:1118, 1960.
2. I. S. Lisker, Inzh.-Fiz. Zhurn. (3), 1962.

CALCULATION OF THE CONDITIONS FOR PULLING FROM THE MELT SEMICONDUCTOR SAMPLES WITH A GIVEN IMPURITY DISTRIBUTION ALONG THEIR LENGTH

B. M. Gol'tsman and K'ung Kuang-lin

The present communication deals with a program for pulling from the melt samples for making thermoelements with a given impurity distribution along their length. This impurity distribution is obtained by varying the effective segregation (distribution) coefficient or by varying the rate of melt feed. Special cases of linear and periodic distributions are considered.

The development of thermoelectric devices made from semiconductors has created the need for growing single crystals and polycrystals with directional properties and a given distribution of composition of the base and the alloying element along the sample length. If the base is a binary alloy (two intermetallic compounds) with the components forming a continuous series of solid solutions, then to reduce the lattice thermal conductivity of the sample and to increase its thermoelectric power it is necessary to increase the concentration of the component with the larger energy gap in the direction from the cold to the hot junction [1]. The carrier density n (the concentration of the alloying component) should also rise along the same direction.

The theory of the problem has two aspects:

1. Calculation of the optimum distribution of composition in the sample.
2. Calculation of a crystallization program ensuring the required composition distribution.

Conditions are considered here for ensuring the required impurity distribution $C(l)$ along the sample length in the case of pulling from the melt by the Czochralski method. The following ways of controlling the crystallization method are discussed:

(a) Variation of the effective segregation coefficient k (e.s.c.) by varying the rate v of pulling of the crystal or by altering the mixing conditions in the melt.

(b) Variation of the rate at which the impurity enters the melt. This is done by varying the rate v_f of feeding a uniform rod into the melt or by using a feed rod with the impurity concentration C_f varying along its length.

The problem is to determine the dependences $k = k(l)$, $v_f = v_f(l)$, or $C_f = C_f(l)$ corresponding to a specified distribution $C = C(l)$ of the impurity in the sample.

Let us formulate the initial assumptions made to simplify the problem: 1) we shall consider crystallization of a material consisting of two components, the base and the alloying admixture; 2) the concentration of the alloying admixture is small and therefore the e.s.c. is independent of the impurity concentration in the melt and the solid; 3) the impurity is distributed uniformly across the sample cross section; 4) the parameters k, C_f, or v_f vary sufficiently slowly for the impurity distribution in the melt to be stationary; 5) the density ρ of the melt and of the crystal is the same; 6) the cross section S of the crystal and the cross section S_f of the feed rod are constant; 7) the amount A of the impurity evaporating from the whole melt surface per unit time is proportional to its concentration C_f in the melt: $A = \alpha C_f$, where α is a constant.

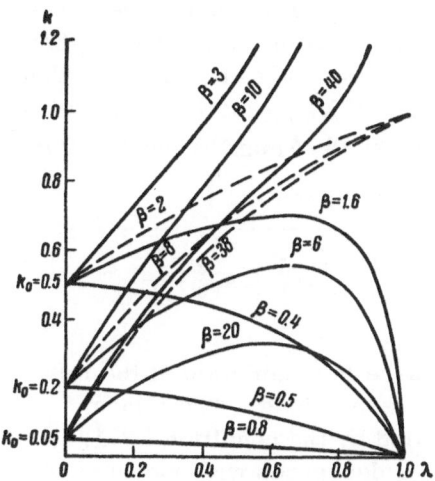

Fig. 1. Dependence of the effective segregation coefficient on the fraction of the melt which has solidified in the case of a linear impurity distribution.

The equation relating the function $C(l)$ and its first derivative with respect to l to the crystallization parameters has the form

$$\frac{C'}{C} - \frac{k'}{k}$$
$$= \frac{1}{\rho V}\left[\rho S(1-k) + \frac{v_f}{v}\rho_f S_f\left(\frac{C_f}{C}k - 1\right) - \frac{a}{v}\left(1 - \frac{C}{k}\right)\right], \quad (1)$$

where V is the volume of the melt at a given moment; ρ_f is the density of the feed rod.

The primes in Eq. (1) denote differentiation with respect to l. It is difficult to analyze Eq. (1) in its general form and therefore we shall consider special cases when one of the parameters k, v_f, or C_f is constant.

Variation of the e.s.c. No Evaporation or Feeding

Integrating Eq. (1) and taking $\alpha = v_f = 0$, we find

$$\frac{1}{k} = \frac{C_0}{C}\left(\frac{1}{1-\lambda}\right)\left(\frac{1}{k_0} - \int_0^\lambda \frac{C}{C_0}\,d\lambda\right). \quad (2)$$

The subscript 0 refers to quantities taken at the initial moment of crystallization; $\lambda = lS/V_0$ is the proportion of the melt which has solidified.

We shall determine the crystallization program for a given $C(\lambda)$ law by means of Eq. (2) and by using the dependence of k on the rate of growth and the mixing conditions. This dependence can be calculated [2] or found empirically. We shall consider some special cases.

a) Uniform Impurity Distribution $C = C_0 = $ const. From Eq. (2) we have

$$k = k_0\frac{1-\lambda}{1-k\lambda}.$$

We shall not consider this case in more detail because it has been analyzed in published work [3, 4].

b) Linear Distribution of the Impurity along the Crystal Length. Let the distribution function have the form $C = C_0(1 + bl)$. Let $\beta = bV_0/S$; then $C = C_0(1 + \beta\lambda)$. Substituting the distribution function into Eq. (1) and integrating we obtain

$$k = \frac{(1 + \beta\lambda)(1 - \lambda)}{\frac{1}{k_0} - \lambda - \frac{\beta}{2}\lambda^2}. \quad (3)$$

We shall consider the case $k_0 < 1$. Let k vary from k^* to 1 (k^* is the equilibrium segregation coefficient). Figure 1 gives the dependences $k = k(\lambda)$ plotted using Eq. (3) for various values of β and k_0. When $\beta < (1 - k_0)$, the value of k decreases monotonically to 0 with increase of λ to 1. When $\beta > (1 - k_0)$ the value of k first increases with increase of λ, and then decreases; if $\beta < 2(1-k_0)/k_0$ then $k = k_{max} < 1$. If $\beta = 2(1 - k_0)/k_0$, then $k_{max} = 1$ when $\lambda = 1$. If $\beta > 2(1 - k_0)/k_0$, then $k_{max} = 1$ when $\lambda < 1$.

This is explained as follows: When $\beta = 2(1 - k_0)/k_0$, the slope of the straight line $C(\lambda)$ is such that the initial amount of the impurity in the melt is distributed linearly along the whole crystal. For lower values of β there is an impurity excess in the melt at the end of the crys-

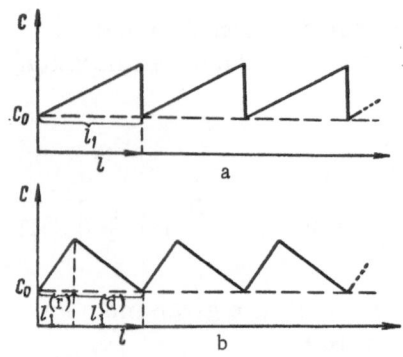

Fig. 2. "Linear-periodic" impurity distribution.

tallization process and the e.s.c. should decrease. The
value of λ_{max} for the "linear" part of the crystal is then
limited by the value k = k*. From Fig. 1 it is clear that
when $\beta < 2(1 - k_0)/k_0$ the value of λ_{max} increases with increase of β. We find the value of λ_{max} by substituting
k = k* into Eq. (3):

$$\lambda_{max} = \frac{k^* + \beta - 1}{(2 - k^*)\beta} + \sqrt{\left[\frac{k^* + \beta - 1}{(2 - k^*)\beta}\right]^2 + \frac{2}{\beta} \cdot \frac{1 - \frac{k^*}{k_0}}{2 - k^*}}.$$

When $\beta > 2(1 - k_0)/k_0$ there is a depletion of the impurity in the melt and the e.s.c. rises. The "linear"
portion in that case is limited by the condition k = 1. It
is clear from Fig. 1 that when $\beta > 2(1 - k_0)/k_0$ the value of λ_{max} decreases with increase of
β. Taking k = 1 in Eq. (3) we find

$$\lambda_{max} = 1 - \sqrt{1 - \frac{2}{\beta k_0}(1 - k_0)}.$$

In the same way, Eq. (3) may be analyzed for the case when k varies not from k* to
unity but in a narrower interval and for the case when $k_0 > 1$.

c) Periodic Variation of the Impurity Concentration along the Sample Length. Samples
used for thermoelements are usually short and a large number of such samples can be cut
from one crystal. Therefore it is interesting to study a periodic variation of the impurity concentration along the crystal length.

We shall consider a "sawtooth" impurity distribution in which each cycle consists of a
linear "rise" of the concentration and a "decay" region; for the sake of simplicity we shall assume that the "decay" region is negligible in extent compared with the "rise" region (Fig. 2a).
Let the concentration in the sample rise in each cycle according to the law $C = C_0(1 + bl)$, where
l is the coordinate taken from the beginning of the cycle. The dependence of the e.s.c. k_{n+1},
corresponding to the (n +1)th cycle, on l is given by the expression

$$k_{n+1} = \frac{(1 + bl)(l_0 - nl_1 - l)}{\frac{l_0}{k_0} - nl_1 - \frac{b}{2} nl_1^2 - l - \frac{b}{2} l^2},$$

(4)

where l_1 is the length of the cycle, and k_0 is the e.s.c. at the beginning of growth.

We shall find the dimensions of the region of the crystal in which such a "linear-periodic"
impurity distribution can be obtained. For this purpose we shall write the expression for the
melt concentration in the (n+1)th cycle

$$C_{L,\,(n+1)} = \frac{C_0(1 + bl)}{k_n + 1} = C_0 \frac{\frac{l_0}{k_0} - nl_1 - \frac{b}{2} nl_1^2 - l - \frac{b}{2} l^2}{l_0 - nl_1 - l}$$

and find how $C_{L,\,(n+1)}$ varies as a function of n. For simplicity we shall assume that the volume of the crystal in the region corresponding to one cycle is much smaller than the volume of
the melt, i.e., that $C_{L,\,(n+1)}$ is constant within one cycle and equal to its value at $l = 0$. Then

$$C_{L,\,(n+1)} = C_0 \cdot \frac{\frac{l_0}{k_0} - nl_1 - \frac{b}{2} nl_1^2}{l_0 - nl_1}.$$

(5)

From formula (5) it follows that $C_{L,(n+1)}$ is independent of n if the condition $bl_1 = 2(1-k_0)/k_0$ is satisfied. For such a condition the average impurity concentration in the cycle is equal to the concentration in the melt at the beginning of the cycle and therefore the impurity concentration should be the same for all the cycles. In such a case we should have a "linear-periodic" impurity distribution along the whole length of the crystal.

Henceforth we shall assume that the e.s.c. varies from k^* to 1, where $k^* < 1$. If $bl_1 < 2(1 - k_0)/k_0$, then as n increases, the value of $C_{L,n}$ increases and k_n decreases. The maximum possible number of cycles is found by substituting $k_n = k^*$ into Eq. (4) and assuming that the minimum value of k is obtained when $l = 0$:

$$n_{max} = \frac{l_0}{l_1} \cdot \frac{\frac{1}{k^*} - \frac{1}{k_0}}{\frac{1}{k^*} - 1 - \frac{b}{2} l_1} + 1. \tag{6}$$

If $bl_1 > 2(1 - k_0)/k_0$, then on increase of n the value of $C_{L,n}$ decreases and k_n increases. The maximum number of cycles is found by substituting $k_n = 1$ with $l = l_1$ in Eq. (4):

$$n_{max} = \frac{l_0}{l_1} \cdot \frac{2\left(1 + bl_1 - \frac{1}{k_0}\right)}{bl_1}. \tag{7}$$

In the values of n_{max} found by means of Eqs. (6) and (7), the fractional part should be rejected and only the integer kept. In the same way we can easily determine n_{max} when $k > 1$.

d) Arbitrary Variation of the Impurity Concentration along the Sample. We shall represent the function $C(\lambda)$ in the form of a series in powers of λ:

$$C(\lambda) = \sum_{n=0}^{\infty} \gamma_n \lambda^n. \tag{8}$$

Substituting Eq. (8) into Eq. (1) and integrating, we obtain

$$k = \frac{(1 - \lambda) \sum_{n=0}^{\infty} \gamma_n \lambda^n}{\frac{C_0}{k_0} - \sum_{n=0}^{\infty} \frac{\gamma_n}{n+1} \lambda^{n+1}}. \tag{9}$$

It can easily be shown that if the series (8) converges the series in the denominator of Eq. (9) also converges.

Variation of the e.s.c. Evaporation and Feeding Taken into Account

The problem consists of integrating Eq. (1) in its general form. However, this equation cannot in general be solved, since in addition to k it contains v and therefore it is necessary to know the dependence k(v) for each particular case.

One possible case when feeding can easily be allowed for is the growth of a crystal with a "periodic" impurity distribution (provided we can neglect the variation in the melt composition during one cycle). Then the required impurity distribution in the crystal within one cycle is obtained by varying the e.s.c. and keeping the composition and volume of the melt constant by the feed.

Let $C = C(l)$ be the distribution function of the impurity concentration in one cycle (l is taken from the beginning of the cycle). From the constancy of the volume and composition of the melt we obtain formulas which give the cross section, concentration, and rate of immersion of the feed rod in the melt

$$S_f l_f = S l_1,$$

$$C_f = \frac{1}{l_1} \int_0^{l_1} C(l)\, dl,$$

where l_f is the part of the rod which enters the melt during one cycle.

Variation of the Rate of Feed. Constant Value of the e.s.c.

We shall consider two cases: 1) the rate of feed v_f of the rod to the melt is constant, and the impurity concentration C_f in the rod varies along its length; 2) the feed rod is uniform, and the rate at which the rod enters the melt varies.

Case 1 (v_f = const). We shall select v_f so that the volume of the melt is constant. This leads to the condition $S_f v_f$ = Sv. Then from Eq. (1) we obtain

$$C_f = \frac{\rho}{\rho_f} \frac{V}{S} \frac{C'}{k} + \frac{C}{k} - \frac{\rho}{\rho_f} C\left(\frac{1}{k} - 1\right) + \frac{\alpha C}{\rho_f S k v} . \tag{10}$$

If the concentration C_f is small, i.e., $\rho_f = \rho$, then Eq. (10) is simplified:

$$C_f = \frac{V C'}{\rho k} + C + \frac{\alpha C}{\rho S k v} . \tag{11}$$

We shall consider the "linear-periodic" impurity distribution shown in Fig. 2b. The quantities referring to the "rise" and "decay" regions will be denoted by the superscripts (r) and (d). The lengths of the portions corresponding to "rise" and "decay" will be denoted by $l_1^{(r)}$ and $l_1^{(d)}$. The distribution function of the impurity in the "rise" regions will be written in the form

$$C^{(r)} = C_0 (1 + bl).$$

The coordinate l is taken from the origin of the region considered. In the "decay" regions the distribution function is

$$C^{(d)} = C_0 \left[(1 + b l_1^{(r)}) - bl \cdot \frac{l_1^{(r)}}{l_1^{(d)}} \right].$$

To simplify the formulas, evaporation of the impurity will be neglected. From Eq. (11) we find the function $C_f(l)$ for the "rise" and "decay" regions:

$$C_f^{(r)} = C_0 \left(1 + bl + \frac{b l_0}{k} \right),$$

$$C_f^{(d)} = C_0 \left[1 + b l_1^{(r)} - \frac{b l_1^{(r)} l_0}{l_1^{(d)} k} - \frac{b l_1^{(r)}}{l_1^{(d)}} l \right],$$

where $l_0 = V_0/S$.

The quantities $C_f^{(r)}$ and $C_f^{(d)}$ should be positive for all values of l. Let $b > 0$. The concentration $C_f^{(r)}$ is positive for any values of l_0 and k and for any positive value of b. The concentration $C_f^{(d)}$ is positive for all values of l if the following inequality is satisfied:

$$l_1^{(d)} \geqslant \frac{b l_1^{(r)} l_0}{k} . \tag{12}$$

In order to be able to use the "rise" as well as the "decay" regions in a cut crystal, the lengths of these regions should be equal. Then the inequality (12) leads to a condition which limits the slope of the $C(l)$ lines:

$$b \leqslant \frac{k}{l_0}.$$ (12a)

If the condition (12a) is not satisfied and only the "rise" regions are used, then to reduce the lengths of the "decay" portions we can feed the melt with the base material at the times corresponding to the "decay" portions. In this case the lengths of the "decay" regions are found by integrating Eq. (11) with $\alpha = C_f = 0$:

$$l_1^{(d)} = \frac{l_0}{k} \ln (1 + b l_1^{(r)}).$$

<u>Case 2</u> (C_f = const). Equation (1) is integrated by quadrature, assuming that we can neglect the change in the melt volume due to impurity evaporation. However, the solution of the equation is very cumbersome and unsuitable for practical use. It is simplified if we can neglect the change in the melt volume due to the feed (the melt is fed with a material strongly enriched with the impurity). Then, solving Eq. (1) we obtain

$$v_f = \frac{1}{\rho_f S_f C_f} \left[\frac{C'\rho v}{k} (V_0 - Sl) - \frac{1-k}{k} C\rho v S - \frac{\alpha C}{k} \right],$$

where v_f is a positive quantity at all values of l. This leads to the condition

$$\frac{d(\ln C)}{dl} \geqslant \frac{1-k}{l_0 - l} + \frac{\alpha}{\rho v S (l_0 - l)}.$$ (13)

If we neglect the impurity evaporation, then the condition (13) is written simply as

$$\frac{d(\ln C)}{d\lambda} \geqslant \frac{1-k}{1-\lambda}.$$ (14)

We shall consider a linear impurity distribution in the crystal $C = C_0(1 + \beta\lambda)$, $\beta > 0$. Substituting the distribution function into the inequality (14) we obtain

$$\frac{\beta}{1 + \beta\lambda} \geqslant \frac{1-k}{1-\lambda}.$$ (15)

Let k < 1. Both parts of the inequality (15) are positive for any λ. The left-hand part of the inequality decreases on increase of λ, but the right-hand part increases. Consequently the inequality (15) should be satisfied first of all when $\lambda = 0$, i.e.,

$$\beta \geqslant (1 - k).$$ (16)

However, when the condition (16) is satisfied and k < 1, it is impossible to obtain a linear distribution along the whole length of the sample by means of feeding. In fact, at values $\lambda < 1$ the left-hand part of the inequality (15) becomes less than unity and the right-hand part greater than unity, i.e., the inequality is not satisfied. This is so because the rate of enrichment of the melt with impurity as a result of crystallization increases without limit with a decrease of the melt volume, while the concentration of the impurity in the crystal increases linearly. We find the value of λ_{max} for a region with a linear impurity distribution by equating the left- and right-hand parts of Eq. (15):

$$\lambda_{max} = \frac{\beta - 1 + k}{\beta(2 - k)}.$$

If k < 1, then when $\beta > 0$ a linear impurity distribution is obtained along the whole length of the crystal.

LITERATURE CITED

1. F. D. Rosi, E. F. Hockings, and N. E. Lindenblad, RCA Rev. 22:82, 1961.
2. I. Burton, R. Prim, and W. Slichter, J. Chem. Phys. 21:1987, 1953.
3. Yu. M. Shashkov, Metallurgy of Semiconductors, Metallurgizdat, 1960.
4. I. Burton, E. Kolb, W. Slichter, and I. Struthers, J. Chem. Phys. 21:1991, 1953.

THERMOELECTRIC PROPERTIES OF SILVER-DOPED CdSb

I. M. Pilat and L. I. Anatychuk

Impurities exert a considerable influence on the physical properties of cadmium antimonide [1-3]. By introducing an impurity we can change the type of conduction and obtain n- or p-type samples. Experimental investigations of the electrical and galvanomagnetic properties have established [3] that the type of conduction of impure samples is governed by the simple valence rule, as in the case of elemental semiconductors of group IV (Ge, Si) or $A^{III}B^{V}$ compounds.

The influence of elements of group I, in particular silver, has not yet been studied sufficiently. Justi and Lautz [4] investigated the influence of a silver impurity on the Wiedemann–Franz–Lorentz constant. Turner et al. [5] found that samples with a silver impurity and a hole density of $3 \cdot 10^{18}$ cm^{-3} had a thermoelectric power of 323 μV/deg and a thermal conductivity of 10^{-2} W $\cdot$ cm$^{-1} \cdot$ deg^{-1}.

In the work reported here, we investigated the temperature dependence of the electrical conductivity σ, of the thermoelectric power α, of the Hall coefficient R, and of the thermal conductivity κ in the temperature interval 90-420°K. The thermal conductivity was measured by the absolute method using apparatus similar to that described in [6]. The heat losses due to radiation and due to conduction along thermocouples and heater leads were determined experimentally and allowed for in the calculation of the thermal conductivity. The study was carried out on polycrystalline samples containing various amounts of silver.

The CdSb samples were prepared by fusing antimony of grade SU-000 and cadmium of grade Kd-0, the latter having been subjected to quadruple fractional vacuum distillation followed by zone-melting purification. The alloys were subjected to additional recrystallization (18 passes) and were then doped with silver (samples 2-5). The table lists the compositions and properties of the alloys at room temperature.

As expected, the silver impurity did not alter the type of conduction in CdSb. The Hall coefficient and the thermoelectric power remained positive for all samples at all temperatures in the investigated interval. Figure 1 gives the temperature dependences of the thermoelec-

Sample No.	Composition	σ, ohm$^{-1} \cdot$cm^{-1}	α, μV/deg	$\kappa \cdot 10^3$, cal $\cdot$cm$^{-1} \cdot$sec^{-1} $\cdot$deg^{-1}	R, cm^3/C	N$\cdot 10^{-16}$,cm^{-3}
1	CdSb	0.6	395	2.6	125	2.08
2	CdSb + 0.01% Ag	4.1	395	2.8	13	29
3	CdSb + 0.1% Ag	85.79	280	3.15	0.85	630
4	CdSb + 1% Ag	147.6	235	4.0	0.32	$2 \cdot 10^3$
5	CdSb + 3% Ag	1818	72	7.5	—	—

*Carrier densities are given for 90°K.

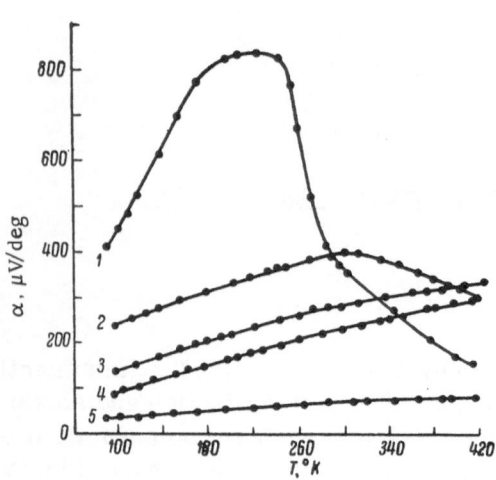

Fig. 1. Temperature dependence of the thermo-electric power. The numbers of the samples are the same as in the table.

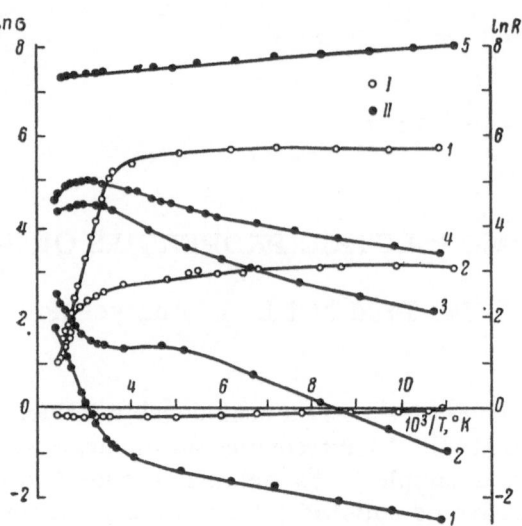

Fig. 2. Temperature dependences of the electrical conductivity and the Hall coefficient. I) Hall coefficient; II) electrical conductivity. The curves are numbered as in Fig. 1.

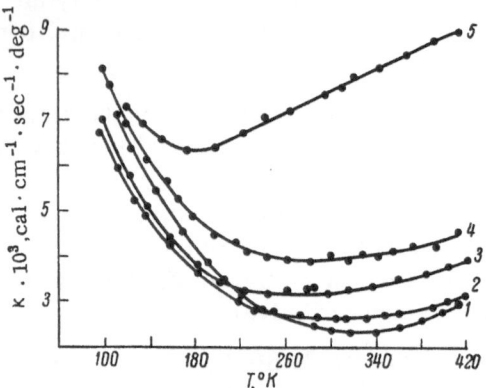

Fig. 3. Temperature dependence of the thermal conductivity. The curves are numbered as in Fig. 1.

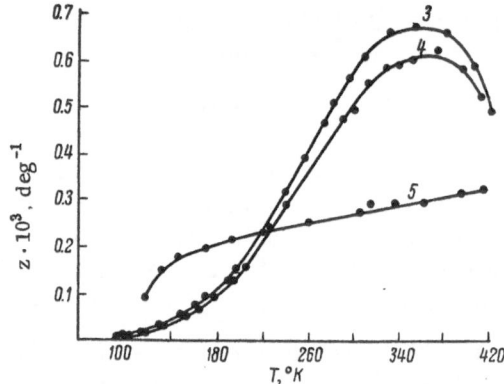

Fig. 4. Temperature dependence of the thermo-electric efficiency. The curves are numbered as in Fig. 1.

tric power. The power decreases with increase of the impurity concentration and the maximum of α exhibited by stoichiometric samples disappears.

The temperature dependences of the electrical conductivity and the Hall effect (Fig. 2) were typical of impure semiconductors. A region with a positive temperature coefficient appeared on doping. With increase of the impurity concentration, this region broadened and in the case of sample No. 5 was observed throughout the investigated temperature range. The Hall effect measurements indicated that the silver impurity dissolves quite well in CdSb, since the change in the carrier density largely matched the impurity concentration.

The temperature dependence of the thermal conductivity of the test is given in Fig. 3.

It should be noted that the silver-doped samples had a relatively high thermoelectric efficiency (figure of merit) z. Quite high values of z were obtained for samples with 0.1 and 1% silver. In particular, $z = 0.56 \cdot 10^{-3}$ deg^{-1} for sample No. 3 at 300°K, reaching a maximum of $0.67 \cdot 10^{-3}$ deg^{-1} at 362°K. The temperature dependences of the efficiency are given in Fig. 4.

The measured values of z indicate that cadmium antimonide may be used to manufacture positive branches of thermoelements if it is suitably doped, since the introduction of the silver impurity alone gave values of z of the same order as for ZnSb [7], which is currently used in the manufacture of thermoelectric generators.

LITERATURE CITED

1. I. M. Pilat, Fiz. Metal. i Metalloved 4:232, 1957.
2. I. M. Pilat, V. D. Iskra, and V. B. Shuman, Fiz. Tverd. Tela 1:393, 1959.
3. I. M. Pilat, in: Problems of Metallurgy and Physics of Semiconductors, Izd. Akad. Nauk SSSR, 1961, p. 81.
4. E. Justi and G. Lautz, Z. Naturforsch. 7a:191, 1952.
5. W. J. Turner, A. S. Fishler, and W. E. Reese, Phys. Rev. 121:759, 1961.
6. E. D. Devyatkova and I. A. Smirnov, Zhur. Tekh. Fiz. 27:1944, 1957.
7. A. F. Ioffe, Semiconductor Thermoelements, Izd. Akad. Nauk SSSR, 1960.

SOME PROPERTIES OF DOPED CADMIUM ANTIMONIDE

S. M. Gusev and G. V. Rakin

Cadmium antimonide has been investigated by many workers [1-6]. At room temperature its electrical conductivity is $\sigma \approx 0.5 \ \text{ohm}^{-1} \cdot \text{cm}^{-1}$, its thermoelectric power is $\alpha \approx 100 \ \mu\text{V/deg}$, and its thermal conductivity is $\kappa \approx 1 \cdot 10^{-2} \text{W} \cdot \text{cm}^{-1} \cdot \text{deg}^{-1}$. The low values of σ and α at room temperature mean that sufficiently high values of the thermoelectric efficiency z cannot be obtained. Therefore we made an attempt to obtain materials with z higher than that of stoichiometric CdSb by means of doping. The electrical conductivity, thermoelectric power, and thermal conductivity were investigated for cadmium antimonide of stoichiometric composition and doped with Cu, Ga, In, Ge, Sn, Se, Te. For some of the samples, the values of the thermoelectric efficiency were determined. The CdSb single crystals, prepared by zone recrystallization, had a carrier density of 10^{15}cm^{-3}.

From the temperature dependence of the electrical conductivity (Fig. 1) it is clear that doping with copper, germanium, and tin up to 1% increases the conductivity to 500 $\text{ohm}^{-1} \cdot \text{cm}^{-1}$ and raises the carrier density. This may indicate that the solubility of Cu, Ge, and Sn in CdSb is considerable. Doping with small amounts of Ga, In, Se, and Te raises the electrical conductivity at room temperature: further increase of the impurity content does not increase greatly the conductivity, indicating a limited solubility of these elements in CdSb.

It follows from the temperature dependence of the thermoelectric power (Fig. 2) that increase of the percentage content of Cu, Ge, and Sn increases α at room temperature but decreases it at low temperatures, the sign of the power remaining positive at all the test temperatures. When the concentration of these elements is greater than 1%, the thermoelectric power remains constant (about 200 $\mu\text{V/deg}$).

Near room temperature, In, Ga, Se, and Te alter the sign of the thermoelectric power to negative ($\alpha \approx 400 \ \mu\text{V/deg}$). From our experimental data we may conclude that elements of groups I and IV in Mendeleev's periodic table are acceptors in CdSb; if substitution occurs, then Cu atoms replace Cd atoms, and Ge and Sn atoms replace Sb in CdSb with an electron deficiency. Elements of groups III and VI are donors; Ga and In replace Cd, while Se and Te may replace Sb atoms having excess electrons.

To determine z an attempt was made to estimate the thermal conductivity κ at room temperature for the alloyed samples and as a function of temperature for stoichiometric CdSb; the comparison method was used in the temperature range 150-400°K (Fig. 3). In the impurity region the thermal conductivity increases with temperature, indicating phonon scattering; above room temperature the value of κ depends weakly on temperature. At room temperature $\kappa = 2.1 \cdot 10^{-2} \text{W} \cdot \text{cm}^{-1} \cdot \text{deg}^{-1}$. The table lists some data for CdSb with impurities which give the maximum values of z. Ga and Se gave analogous results to those of samples doped with In and Te.

The samples doped with large amounts of selenium and tellurium (Figs. 4 and 5) should be considered separately. On introduction of Se and Te a clearly marked impurity region appears, the slope of which depends on the impurity concentration. From the temperature dependence of the thermoelectric power it follows that the materials obtained have a positive thermoelectric power in the impurity region. Annealing alters considerably the electrical prop-

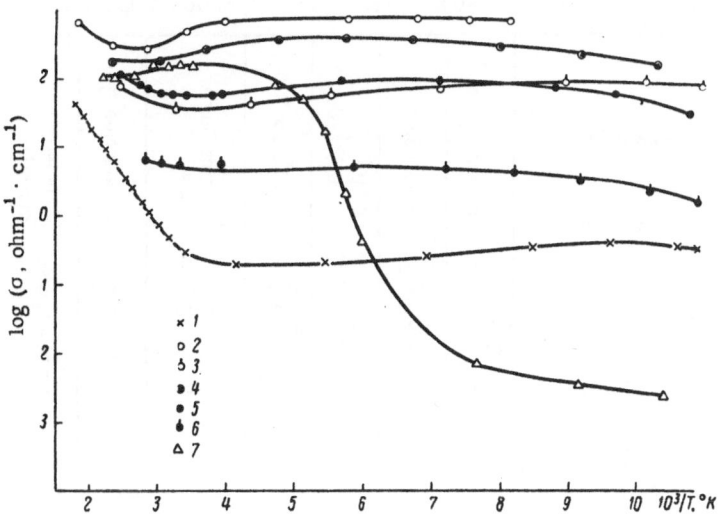

Fig. 1. Temperature dependence of the electrical conductivity of doped CdSb. 1) CdSb single crystal; 2) CdSb + 1% Cu polycrystal; 3) CdSb +Cu single crystal; 4) CdSb + 1% Sn polycrystal; 5) CdSb + 1% Ge polycrystal; 6) CdSb + Ge single crystal; 7) CdSb + Te single crystal.

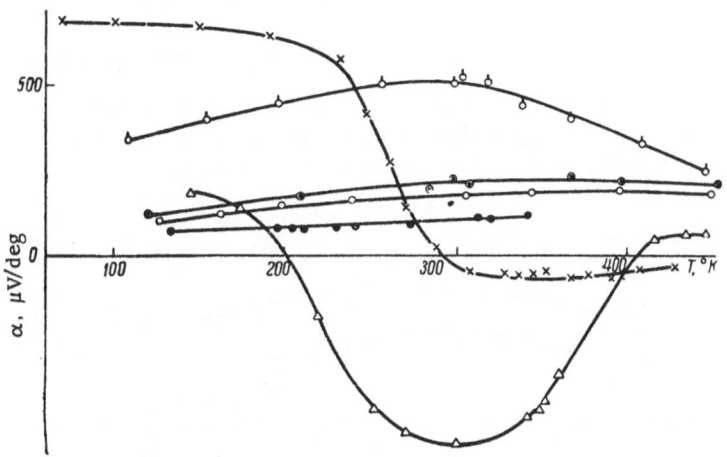

Fig. 2. Temperature dependence of the thermoelectric power. The notation is the same as in Fig. 1.

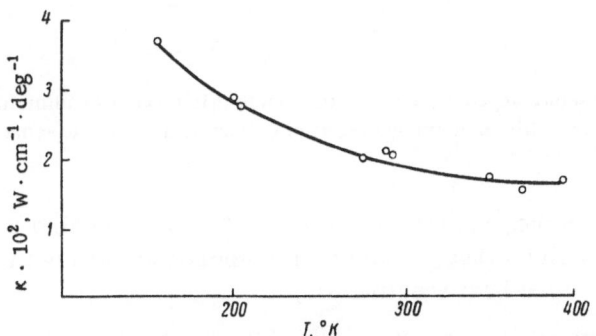

Fig. 3. Temperature dependence of the thermal conductivity of a CdSb single crystal.

Thermoelectric Properties of CdSb Samples Doped with Various Impurities

Composition	σ, ohm$^{-1}\cdot$cm^{-1}	α, μV/deg	$\kappa\cdot10^2$, W$\cdot$cm$^{-1}\cdot$deg^{-1}	$z\cdot10^3$, deg^{-1}	Carrier density, cm^{-3}	Conduction type	Structure of crystals
CdSb	0.5	100	2.1	—	$2\cdot10^{15}$	p	s
CdSb + 1% Cu	484	172	1.83	0.78	—	p	p
CdSb + Cu	32	500	1.9	0.405	$3.6\cdot10^{17}$	p	s
CdSb + 1% Sn	136	224	2.5	0.68	—	p	p
CdSb + 1% Ge	278	120	1.9	0.21	—	p	p
CdSb + Ge	6.34	500	2	0.079	—	p	s
CdSb + Te	20	400	2	0.16	—	n	s
CdSb + 2.67 Te	10	230	1.09	0.047	—	n	p

Note: s denotes a single crystal, and p a polycrystal.

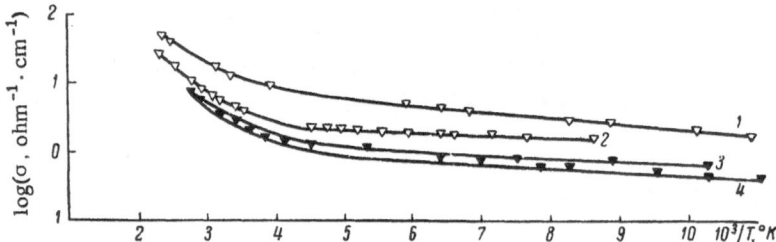

Fig. 4. Temperature dependence of the electrical conductivity of cadmium antimonide doped with large amounts of Se and Te. 1) CdSb + 4.67% Te; 2) CdSb + 2.67% Te; 3) CdSb + 3.25% Se; 4) CdSb + 1.66% Se.

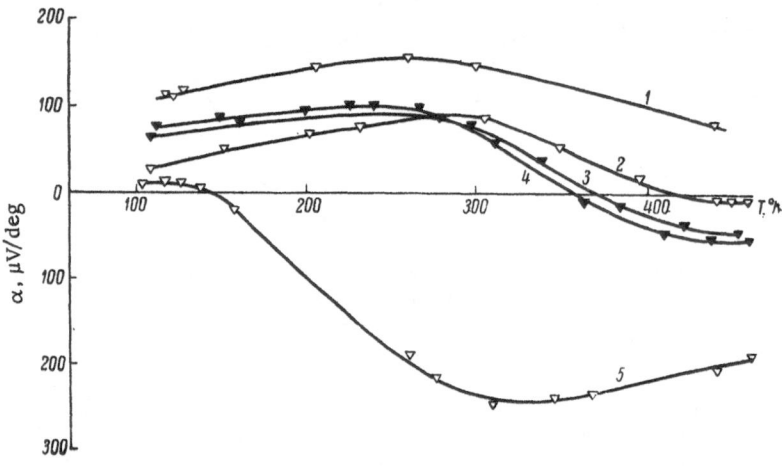

Fig. 5. Temperature dependence of the thermoelectric power of cadmium antimonide doped with selenium and tellurium. The notation is the same as in Fig. 4.

erties of CdSb; thus, for example, sample No. 5 (Fig. 5) was annealed at 350°C for 30 hr and cooled slowly in 24 hr; it is clear that the value of α increases at high temperatures (the other curves represent samples before heat treatment).

Moreover, the introduction of more than 1% of Se and Te increases the melting point to above 600°C (CdSb melts at 456°C); further increase of the Se and Te content in CdSb produces an even stronger rise of the melting point. Preliminary results of a microanalysis indicate that two-phase regions occur in samples with high concentrations of Se and Te. X-ray structure analysis confirms the presence of CdTe and CdSe.

It is not clear why relatively small concentrations of Se and Te increase the melting point considerably.

CdSb doped with Cu, Ge, Sn may be used to prepare positive branches of thermoelements; when doped with In, Ga and Se, Te it may be used for negative branches. Moreover, CdSb+Te and CdSb+Se should be studied in greater detail in connection with obtaining semiconductors with a higher melting point.

LITERATURE CITED

1. E. Justi and G. Lautz, Z. Naturforsch. 7a:602, 1952.
2. M. V. Kot and I. K. Andronik, Uchenye Zapiski Kishenevsk. Gosudarst. Univ., Ser. Fiz.-Mat. Nauk 24:209, 1957.
3. I. K. Andronik, Uchenye Zapiski Kishenevsk. Gosudarst. Univ., Ser. Fiz.-Mat. Nauk, 24:215, 1957.
4. W. J. Turner, A. S. Fischler, and W. E. Reese, Phys. Rev. 121(3):759, 1961.
5. H. Hida and K. Daigaku, R. J. Phys. 5:3654, 1960.
6. I. M. Pilat, Dissertation for Candidate's Degree, Chernovtsy, Gosudarst. Univ., 1958.

THERMOELECTRIC PROPERTIES OF THE HIGHER
MANGANESE SILICIDE

V. A. Korshunov and P. V. Gel'd

Work carried out in the Soviet Union on the initiative and under the leadership of Academician A. F. Ioffe has shown that the power applications of thermoelectricity are not only of great scientific interest but also have a real chance of practical realization.

The problem of power applications may be solved only by theoretical and experimental investigations of many aspects. Of special importance is the study of the thermoelectric properties of a wider range of semiconductors than has been investigated at present, in order to produce more efficient materials for thermoelements and to understand more deeply the main relationships governing the electrical and thermal properties of solids.

Practical applications require the development and study of, first of all, refractory materials for thermoelectric generators, because the efficiencies acceptable in industrial use may be obtained only with relatively high temperature differences between the hot and cold ends of the thermoelement. In particular, the production of high-temperature p-type thermoelectric materials is of special importance because they are necessary for use as the positive branches of vacuum and solid thermoelements [1].

In connection with this aspect the compounds of the transition elements have special interest, particularly transition-metal silicides, which frequently have semiconducting properties and high heat stability.

Investigation of the electrical and magnetic properties of transition-element compounds and the establishment of a relationship between these properties and the electron state of the transition-element atoms is also necessary for the understanding of conduction and the nature of the chemical bond in such compounds.

These circumstances induced us to study the electrical properties of silicon—manganese alloys. The temperature dependences of the electrical conductivity and thermoelectric power were investigated, as well as the phase composition of alloys made from vacuum-distilled manganese and single crystals of silicon. The methods of preparing the samples and of investigating their electrical properties (with and without contacts) have been described earlier [2-4].

The results of measuring the electrical conductivity and thermoelectric power of alloys close in composition to the higher manganese silicide are given in Fig. 1. They do not agree with previous data on the phase diagram of the Mn − Si system [5-7], nor with conclusions about the phase composition derived from a study of the electrical properties of alloys in the Mn − Si system [8]. They suggest rather the existence of a compound with variable composition, containing from 46 to 47 wt.% Si.

Metallographic and x-ray structure analyses of the alloys confirm this hypothesis. It was found that in the MnSi − Si system there is only one intermediate phase of variable composition which is stable within the limits $MnSi_{1.67} - MnSi_{1.73}$ (46-47% Si) [9]. This contradicts the established view on the existence of stoichiometric manganese disilicide [5-7] and the existence of Mn_2Si_3.

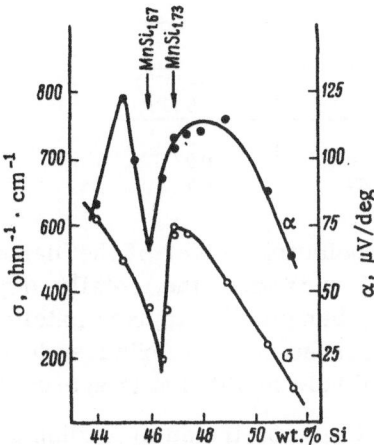

Fig. 1. Dependence of the electrical conductivity and thermoelectric power (with respect to platinum) of MnSi −Si alloys on composition.

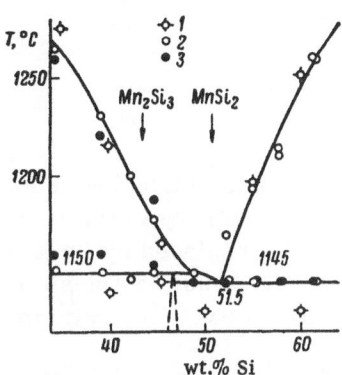

Fig. 2. Phase diagram of the MnSi−Si system. 1) Data of [10]; 2) thermographic data; 3) data obtained by investigating electrical conductivity.

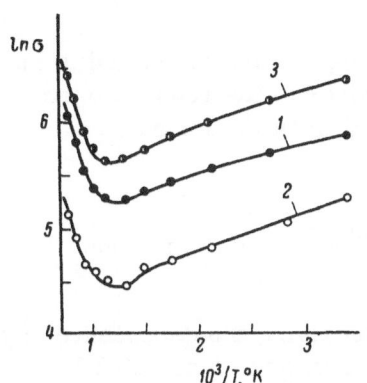

Fig. 3. Temperature dependence of the electrical conductivity of the higher manganese silicide. 1) $MnSi_{1.67}$; 2) $MnSi_{1.70}$; 3) $MnSi_{1.73}$. Curves 1 and 3, $\Delta E_0 = 0.6$ eV; curve 2, $\Delta E_0 = 0.5$ eV.

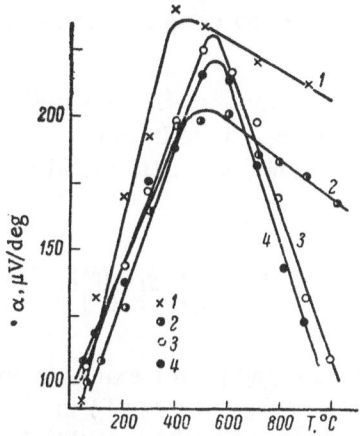

Fig. 4. Temperature dependence of the thermoelectric power of the higher manganese silicide. 1) $MnSi_{1.67}$; 2) $MnSi_{1.70}$; 3) $MnSi_{1.73}$; 4) $MnSi_{1.76}$.

If we start from the assumption that this phase is a solid solution based on the hypothetical $MnSi_2$, we should admit the existence of vacancies (Mn_nSi_{2n-x}) in the silicon sublattice because other possible variants of the structure of the higher silicide which agree with the composition dependence of its stability region (for example, an interstitial or substitutional solid solution of manganese in $MnSi_2$) cannot explain the observed increase of the interplanar spacings with increase of the silicon content.

The data of metallographic and thermal analyses, as well as the results of a study of the temperature dependence of the electrical conductivity of the alloys [3], make it possible to propose an improved variant of the phase diagram of the MnSi − Si system [9] (Fig. 2). The peritectic temperature is close to 1150°C, corresponding to approximately 48.8% Si. The higher manganese silicide forms a eutectic with silicon [10]. The coordinates of the eutectic point are 1145°C and 51.5% Si.

The reported data on the existence of an intermediate phase of variable composition and on the structure of the phase diagram of the MnSi − Si system have recently been confirmed by Dudkin and Kuznetsova [11].

Wt. % Si	Property			
	μ^* (300°K)	n_p, cm^{-3}(300-500°K)	u_p cm$^2 \cdot$V$^{-1}\cdot$sec^{-1}	p ($u \sim T^{-p}$) (300-500°K)
46.5	3.32	$1.2 \cdot 10^{20}$	10	0.9
47.0	2.68	$1.0 \cdot 10^{20}$	40	0.9

Polytherms of the electrical conductivity and thermoelectric power of the higher silicide (Figs. 3 and 4) have, contrary to the results of Nikitin [8], extrema: the metallic dependence at low temperatures is replaced by typical semiconducting behavior at approximately 500°C. At temperatures of 700-1000°C intrinsic conduction is observed and the forbidden band width ΔE_0 amounts to 0.5-0.6 eV. The ratio of the electron and hole mobilities is approximately 4-5.

The existence of an intrinsic conduction region has been confirmed by Dudkin and Kuznetsova [11]. They, however, obtained a value for ΔE_0 which is too low because their measurements were carried out in a narrow range of temperatures covering mainly the transition region.

The present values and the nature of the dependence of the electrical conductivity and thermoelectric power on composition at room temperature are in good agreement with other work [8, 11].

The metallic nature of the conduction near room temperature is related to the carrier (hole) gas degeneracy. This is indicated by the rough values of the reduced chemical potential (μ^*), hole density (n_p), and hole mobility (u_p), calculated from the formulas

$$\alpha = \frac{k}{e}\left[\frac{r+2}{r+1}\cdot\frac{F_{r+1}(\mu^*)}{F_r(\mu^*)} - \mu^*\right], \quad n = 4\pi\left(\frac{2mkT}{h^2}\right)^{3/2}F_{1/2}(\mu^*), \quad \sigma = enu.$$

It is assumed that r (the power exponent in the expression giving the dependence of the carrier path length on the energy, $l = l_0\,\varepsilon^r$) is zero, which corresponds to a covalent lattice and scattering on the acoustic vibrations of this lattice. * Moreover, we assume that the effective carrier mass is equal to the free-electron mass and that it is independent of temperature. †

The table gives values calculated for alloys containing 46.5 and 47.0% Si.

The values of μ^*, n_p, and u_p, as well as the temperature dependence of the mobility, are in agreement with the assumption of hole-gas degeneracy.

The reduction of the density of holes and the increase of their mobility, as well as the increase of the forbidden band width as saturation of the higher silicide with silicon is approached, are obviously related to the gradual filling of vacancies in the silicon sublattice, accompanied by a reduction of the number of defects in the lattice and increase of the unit cell volume.

---

* This hypothesis is supported by the small difference between the electronegativities of manganese and silicon, the approximate data on the magnetic properties of $MnSi_2$ [12, 13], and also by the measured electrical properties of Mn_nSi_{2n-x}. However, the available data do not show conclusively that the binding in Mn_nSi_{2n-x} is purely covalent. Moreover, we must assume that there is some polarization of atoms in the higher manganese silicide.

† It should be noted that in the presence of a narrow "impurity" band the carriers have large effective masses. Therefore the density values given below are somewhat underestimated and the mobilities are overestimated.

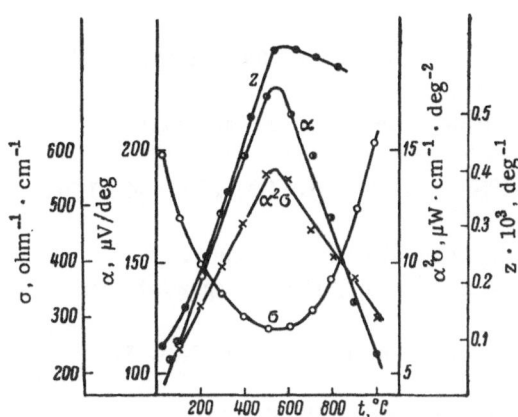

Fig. 5. Temperature dependences of α, σ, $\alpha^2\sigma$, and z for the alloy containing 47% Si (MnSi$_{1.73}$)

As is known [14], there are two possible explanations of metallic conduction in semiconductors. One is based on the possibility of "sucking in" of the local impurity states, for example, into the conduction band; in this case there is a definite number of electrons in the band which is independent of temperature [15]. The other is based on the impurity-band concept [16-18]. Both mechanisms may act together. Obviously, in the case of the higher manganese silicide, with a defective structure in which the vacancies in the silicon sublattice are randomly distributed, one can expect splitting of the energy band or, in other words, splitting of the "impurity" band [19]. The latter circumstance is responsible for the metallic nature of the temperature dependence of the electrical conductivity of Mn$_n$Si$_{2n-x}$ up to 500℃.

The high melting point, heat resistance and stability, and sufficiently large width of the forbidden band of the higher manganese silicide, suggest it as a possible material for thermoelectric generators. It is particularly interesting that the optimum conditions for thermoelements are obtained in this material at 800°K. Figure 5 shows the temperature dependences of α, σ, $\alpha^2\sigma$, and z for the alloy containing 47% Si, i.e., for the higher silicide saturated with silicon. In calculating z we used the experimental values of α and σ, and also the values of the thermal conductivity determined from the formula $\kappa = \kappa_l + \kappa_{el}$, where the lattice conductivity (κ_l) is assumed to decrease in inverse proportion to the absolute temperature and the electronic conductivity (κ_{el}) up to the degeneracy temperature is given by the formula

$$\kappa_{el} = \frac{\pi^2}{3}\left(\frac{k}{e}\right)^2 T\sigma,$$

which is valid in the case of strong degeneracy. The value of the thermal conductivity at room temperature was measured by the method of A. F. Ioffe and was found to be $(10 \pm 2) \cdot 10^{-3}$ cal $\cdot$ cm$^{-1}\cdot$ sec$^{-1}\cdot$ deg^{-1}.

The appropriate calculation of the efficiency of MnSi alloys close in composition to the higher manganese silicide gave the following results for the temperature interval 300-1100°K:*

Wt. % Si	47.0	47.5	45.0
η, %	7	6.5	7.3

It is clear from Fig. 5 that it would be most advantageous to use such alloys as the high-temperature junctions of complex thermoelements. Obviously the preparation and study of alloys or quasi-binary systems based on Mn$_n$Si$_{2n-x}$, which may have even better properties, is of special interest.

In conclusion, it should be noted that the results given above for the efficiency should be checked directly by experimental means. It is also very important to produce refractory materials with similar properties (α, σ, and κ) for the negative branches of thermoelements. Nevertheless, the results described show that the compounds of transition metals, in particular their silicides, are undoubtedly promising materials for thermoelectric generators.

* To calculate z we used the maximum value of $\kappa_{20°C}$. If the average value of the thermal conductivity is taken and it is assumed that this conductivity is independent of temperature, then a calculation of η for MnSi$_{1.73}$ gives 5%.

LITERATURE CITED

1. A. F. Ioffe, Semiconductor Thermoelements, Akad. Nauk SSSR, Moscow-Leningrad, 1960, p. 74.
2. V. A. Korshunov and P. V. Gel'd, Trudy Ural. Politekh. Inst. im. S. M. Kirova, Sverdlovsk, (105):142, 1960.
3. V. A. Korshunov and A. V. Gel'd, Izvest. Vuzov SSSR, Fizika, (6):29, 1960; (4):146, 1961.
4. V. A. Korshunov and P. V. Gel'd, Fiz. Metal. i Metalloved. 11:945, 1961.
5. M. Hansen, Constitution of Binary Alloys [Russian translation], Metallurgizdat, Moscow-Leningrad, 1941.
6. M. Hansen and K. Anderko, Constitution of Binary Alloys, McGraw-Hill, New York – London, 1958.
7. A. S. Berezhnoi, Silicon and Its Binary Systems, Izd. Akad. Nauk UkrSSR, Kiev, 1958. English translation: Consultants Bureau, New York, 1960.
8. E. N. Nikitin, Fiz. Tverd. Tela 1:340, 1959.
9. V. A. Korshunov, F. A. Sidorenko, P. V. Gel'd, and K. N. Davydov, Fiz. Metal. i Metalloved. 12:277, 1961.
10. F. Doerinckel, Z. anorg. u. allgem. Chem. 50:117, 1906.
11. L. D. Dudkin and E. S. Kuznetsova, Doklady Akad. Nauk SSSR 141:94, 1961.
12. G. Foex, J. Phys. 9:37, 1938.
13. Ya. G. Dorfman, Magnetic Properties and Structure of Matter, Gostekhizdat, Moscow, 1955.
14. V. A. Bonch-Bruevich, Fiz. Tverd. Tela, Collection Vol. II, 117, 1959.
15. B. I. Davydov and I. T. Shmushkevich, Uspekhi Fiz. Nauk 24:21, 1940.
16. I. V. Kurchatov, T. N. Kostina, and L. I. Rusinov, Sow. Phys. 7:129, 1935.
17. A. F. Ioffe, Zhur. Tekh. Fiz. 23:431, 1953.
18. S. N. Pekar and M. A. Krivoglaz, Collection dedicated to the memory of S. I. Vavilov, Izd. Akad. Nauk SSSR, Moscow, 1952, p. 334.
19. A. I. Gubanov, Fiz. Tverd. Tela 3:2154, 1961.

INFLUENCE OF ELASTIC VIBRATIONS ON THE PHYSICAL
PROPERTIES OF TERNARY SEMICONDUCTING ALLOYS

A. N. Andreeva, N. N. Pavlov, V. S. Smirnov,
and A. F. Chudnovskii

One of the main problems in the development of semiconductor cooling techniques is that of increasing the efficiency of cooling devices. On the one hand we can develop new semiconducting materials with better electrical properties, and on the other we can try to improve the physical properties of the avilable materials.

It is known that the crystal structure and the mechanical and electrical properties of alloys may be altered by various physical agencies (directional crystallization, rolling, and the irradiation of the melt with elastic waves of various frequencies and intensities). A characteristic feature of the manufacture of thermoelements is the necessity of preparing samples from ternary alloys by compacting. As is known, this is a very difficult operation, producing samples with a wide scatter of properties; it is necessary to use this method because casting gives samples with nonuniform coarse-grained structure, having gas occlusions and poor mechanical and electrical properties.

Earlier work [1-5] has established that elastic waves introduced into the melt ensure the optimum conditions for the formation of fine-grained structure, increase the volume uniformity of the ingot, outgas the melt, and improve the mechanical properties of the alloy.

However, relatively little work has been done on this problem, due principally to the difficulty of transmitting ultrasonic vibrations into the melt at high temperatures. This is particularly complicated in the case of ultrasonic vibrations in melts of semiconducting materials, because their properties are extremely sensitive to the slightest impurity.

Our first experiments dealt with the vibration of the alloy while retaining the available method of melting, using the apparatus shown in Fig. 1. The prepared melt was placed in a quartz ampoule 4, which was evacuated with a fore-pump and sealed. Through a quartz tube 3, attached to the ampoule, and a stainless steel rod 2 the ampoule was connected to an electromagnetic vibrator 1 working at 100 cps. During melting the ampoule was lowered into the lower part of the furnace 5; during crystallization the ampoule was either raised on a stand into the upper part of the furnace (in the case of slow crystallization) or removed from the furnace (in the case of rapid crystallization in air). The vibrator was usually switched on only at the beginning of crystallization for a period of 40 sec to 5 min, depending on the rate of crystallization.

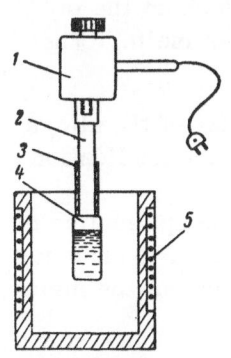

Fig. 1. Vibration apparatus.

One of the main difficulties in crystallization under the action of vibrations is the correct selection of the amplitude and power of the vibrations. The first experiments on vibration melting showed that, if the amplitude is high, the melt appears to boil and splashes onto the ampoule walls. By varying the vibration amplitude we obtained both very porous ingots and ingots which were dense, monolithic, and difficult to fracture. Cylindrical test samples were made from the latter.

To apply ultrasonic vibrations during the crystallization of alloys we used apparatus similar to that described in [4].

The apparatus is shown schematically in Fig. 2. A magnetostriction transducer 1 was connected in an oscillation circuit of 1.5 kW power with a frequency variable from 15 to 25 kc. Transmission of the ultrasonic energy from the vibrator to the melt was achieved by means of a stepped concentrator 2, which, on the one hand, acted as an intermediate link between the vibration source and the melt, and on the other, could be used as an ultrasonic amplifier. The thicker end of the concentrator was rigidly attached to the vibrator and the thinner end was attached to a transmitting rod 3, which entered an aperture in the bottom of the mold 4. The dimensions of the concentrator were designed to make its natural frequency equal to the vibrator frequency. The transmitting rod entered freely the aperture in the mold bottom and was not attached rigidly to the vibrating system. The mold was supported on a special stand.

<u>Method of Investigation.</u> The thermoelectric alloys were prepared by a technique developed at the Institute for Semiconductors of the Academy of Sciences (IPAN) [6, 7].

Two identical ingots 150 g in weight were prepared from the same melt. The ingots were melted in an electrical furnace in graphite crucibles under a layer of graphite powder and heated to 50°C above the melting point. One of the melted ingots was then poured into the graphite mold 4 (Fig. 2) at the moment of switching on the ultrasonic generator. Ultrasonic irradiation of the melt was continued for 30–40 sec until the ingot solidified.

The second (control) ingot melt was crystallized under exactly the same conditions but without ultrasonic irradiation.

Both ingots were then cleaved and the cleavage surface examined for the alloy macrostructure. Then, from the ingot crystallized under the action of ultrasonic waves samples were cut along three directions: vertical, horizontal, and at an angle of 45°. It was not possible to cut regular-shaped samples from the control ingot because it shattered. For comparison we used compacted samples from the irradiated and control materials.

<u>Results of the Investigation.</u> According to the theory of A. F. Ioffe [8], the quality of semiconducting thermoelements is represented by a figure of merit $z = \alpha^2 \sigma / \kappa$, where α, σ, and κ are, respectively, the thermoelectric power, electrical conductivity, and thermal conductivity of the branches.

To determine the influence of ultrasonic vibrations on the electrical properties of alloys and to find whether it would be possible to use case materials to make thermoelements, we measured the thermoelectric power and electrical conductivity of cast and pressed samples of the alloys.

Table 1 lists the averaged results of measurements of the electrical properties of three different alloys.

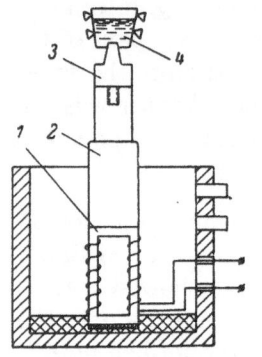

Fig. 2. Ultrasonic apparatus.

It is not yet possible to draw final conclusions about the influence of ultrasonic vibrations on the properties of the melts. The problem is complicated by two factors:

1. The low values of the electrical conductivity of the alloys Nos. 2 and 3 (Table 1) used to select conditions for the application of ultrasonic waves.

2. Oxidation of the alloys on melting before the application of ultrasonic vibrations. No special measures were taken to protect the alloys from oxidation. Later we propose to carry out the melting under a flux or in an inert-gas atmosphere.

However, even from these preliminary data we may draw the conclusion that ultrasonic vibrations not only do not degrade the elec-

60

Table I. Comparison of Sample Properties

Parameter	Positive alloy No. 1		Negative alloy No. 2		Negative alloy No. 3	
	pressed	cast	pressed	cast	pressed	cast
σ, ohm$^{-1}\cdot$cm^{-1}	1400 1500 1350	1600 1500 1450 1550	200 230 230	200 180 180 160	500 580 560	430 560 510 400
α, μV/deg	180 175 175	160 170 165 165	190 180 185	195 200 225 210	200 200 200	230 220 220
$\alpha^2\sigma\cdot10^{-6}$, μV$^2\cdot$deg$^{-2}\cdot$ohm$^{-1}\cdot$cm^{-1}	45 45 45	41 43 39 42	7.2 7.5 7.8	7.5 7.1 9 7	20 23 22	23 24 25 21

Fig. 3. Microstructure of an unirradiated ingot.

Fig. 4. Microstructure of an irradiated ingot.

trical properties of the alloys but make the properties of cast samples approach those of pressed ones.

Macrostructure and Microstructure of the Alloys. The alloy macrostructure was examined on cleaved surfaces of the ingots and showed clearly a reduction in the grain dimensions. The same surfaces were used later to study the alloy microstructure. Moreover, the samples used for microstructure studies were employed for electrical measurements. Figures 3 and 4 show the microstructure of cast unirradiated and irradiated samples of a "negative" alloy (an alloy used for the negative branches). The microstructure of a "positive" alloy was similar.

61

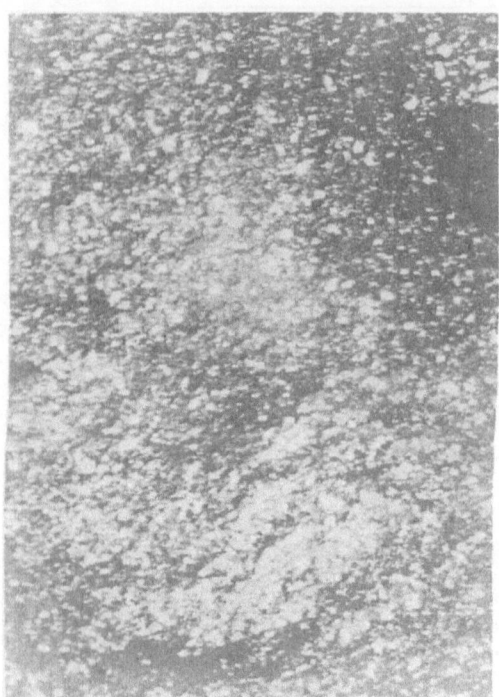

Fig. 5. Macrostructure of an irradiated ingot.

Table II. Measurements of Microhardness (average values from 10 measurements, kg/mm²)

pressed	Samples		
	unirradiated cast	cast after vibration	irradiated cast
Positive alloy			
61.0	19.5	45.8	50.0
38.0	20.6	55.0	61.9
43.8	14.5	58.0	55.0
42.0	23.3	52.5	58.0
42.0	21.3	52.5	60.0
42.0	18.9	55.0	65.8
40.0	20.1	52.5	48.9
42.0	18.9		52.5
42.0	21.3		59.0
42.0	21.9		52.5
Negative alloy			
38.0	17.4	50.0	60.0
37.1	16.0	42.0	65.0
37.5	16.9	55.0	61.0
37.6	21.3	43.8	64.0
37.0	23.3	50.0	50.0
40.2	21.6	45.8	64.0
39.0	20.7	52.5	58.0
42.0	19.5		63.8
42.0	20.7		64.0
42.0	18.4		64.0

Table III

Sample No.	Cast kg/mm²	Sample No.	Pressed irradiated, kg/mm²
14	5.8	20	6.6
19	5.8	21	7.7
30	5.2	25	8.4
32	5.1	27	10.6
35	5.1	3	8.2
3	4.7	5	8.1
		6	9.4

The macrostructure of an irradiated ingot is shown in Fig. 5, where again the reduction in grain dimensions under the action of elastic vibrations can be clearly seen.

The mechanism of the reduction of the grain dimensions by the action of ultrasonic vibrations is related mainly to the development of cavitation processes which break up the growing crystals. Moreover, the following processes play a role: the radiation pressure, which carries crystal fragments back into the interior of the melt where they form new crystallization centers; the intense mixing of the melt, which equalizes the temperature throughout the whole volume and favors volume crystallization. The ultrasound may also affect the nucleation of the primary crystallization centers. The friction forces between the melt and the precipitated needlelike crystals, giving rise to crystal fragmentation and reduction of the grain dimensions, may also be important.

The ultrasonic vibrations also reduce the number of gaseous occlusions and produce a more uniform structure of the material, which raises its plasticity. The more uniform the material as regards grain dimensions, chemical composition, mechanical properties, impurity and occlusion distributions, the higher the plasticity of the material, a fine-grained structure being the most favorable for this state. An increase of the plasticity makes it possible to use the methods of plastic deformation for the production of thermoelements from the alloys subjected to the action of ultrasonic vibrations.

Microhardness of the Alloys. Measurements of the microhardness were carried out using a PMT-3 instrument with a 50-g load and the same samples which were employed for electrical measurements: ingots not subjected to irradiation; ingots solidified under the action of vibrations; ingots irradiated ultrasonically during crystallization; pressed samples. The microhardness was measured at 7-10 points for each sample. The average results of these measurements are listed in Table II.

The following conclusions may be drawn from the microhardness measurements.

1. The unirradiated ingot has the lowest microhardness, which decreases somewhat toward the center of the sample. Some variation of the microhardness along the length of the ingot was also observed.

2. The microhardness of pressed samples was on the average twice as high; those subjected to vibration, 2.5 times greater; and those irradiated, 3 times greater than the microhardness of cast unirradiated samples.

3. Elastic vibrations exert a considerable influence on the ingot uniformity, the values of the microhardness measured along and across the length of the vibration-treated ingot differed by a quantity not exceeding the experimental error.

The data on the compressive strength present the same picture as those on the microhardness. The compressive strength values are given in Table III.

The highest compressive strength was exhibited by the samples prepared by hot pressing from an irradiated alloy.

Conclusions

1. Investigation of the macro- and microstructure of ultrasonically irradiated alloys suggests that it should be possible to prepare ingots with fine-grained structure and with more uniform physical properties than is possible under normal casting conditions.

2. The microhardness tests and measurements of the electrical properties show that ultrasonic irradiation of the melt raises the microhardness and the compressive strength considerably without altering the electrical properties compared with alloys obtained in the normal way.

3. The action of elastic vibrations on the process of crystallization of thermoelectric alloys gave positive results and requires further study.

LITERATURE CITED

1. A. P. Kapustin, Zhur. Tekh. Fiz. 22:765, 1952.
2. A. P. Kapustin, Zhur. Tekh. Fiz. 20:1157, 1950.
3. Applications of Ultrasound in the Manufacture and Heat Treatment of Alloys. Collection of papers, 1-5, Tsentral'noe Pravlenie NTO Mashinostroitel'noi Promyshlennosti, 1961.
4. I. I. Teumin, in: Applications of Ultrasound in Industry, Mashgiz, 1959.
5. V. V. Zaboleev-Zotov and G. I. Pogodin-Alekseev, Metalloved. i Obrabotka Metal. No. 1, 1958.
6. S. S. Sinani and G. V. Kokosh, Fiz. Tverd. Tela 2:1118, 1960.
7. S. S. Sinani, G. N. Gordyakova, and G. V. Kokosh, Zhur. Tekh. Fiz. 28:3, 1958.
8. A. F. Ioffe, Semiconductor Thermoelements, Izd. Akad. Nauk SSSR, Moscow-Leningrad, 1956; English translation: Infosearch Ltd., London, 1957.

MULTISTAGE THERMOELECTRIC GENERATORS [*]

E. K. Iordanishvili and L. S. Stil'bans

According to the theory of A. F. Ioffe [1], the main factor which determines the quality of semiconducting compounds for thermoelectric generators and cooling devices is the figure of merit z:

$$z = \frac{\alpha^2}{\kappa\rho},$$ (1)

where α, ρ, and κ are, respectively, the thermoelectric power, electrical resistivity, and thermal conductivity of the material.

In the case of a thermoelement consisting of positive and negative branches with the optimum cross sections we have

$$z = \frac{(\alpha_p - \alpha_n)^2}{(\sqrt{\kappa_n \rho_n} + \sqrt{\kappa_p \rho_p})^2}.$$ (2)

The quantities α, ρ, and κ_{el} are functions of the carrier density, and therefore there is a value of this density n_0 at which z reaches a maximum. The quantities α, ρ, and κ are also functions of temperature, and ρ depends quite strongly on temperature, varying from $\rho \sim T^a$ ($2.0 \leq a \leq 2.5$) for $Bi_2Te_3 - Sb_2Te_3$ to $\rho \sim T^b$ ($0.8 \leq b \leq 1.2$) for $Bi_2Te_3 - Bi_2Se_3$.

Moreover, the materials currently used have relatively small energy gaps (from 0.12 to 0.3 eV), and this leads, even at temperatures of 150-200°C, to the appearance of effects connected with the presence of carriers of opposite sign. Ambipolar diffusion, which makes a contribution to the thermal conductivity due to heat transport by carrier pairs, plays a particularly important role in the reduction of z with temperature. in materials containing $(1.5-2) \cdot 10^{19}$ cm^{-3} majority carriers, which give maximum z at room temperature, these effects are already noticeable at 350-400°K. All this reduces the value of z of the currently available thermoelectric generators by a factor of several times in the working temperature range. Moreover, the results actually obtained for thermoelectric generators are considerably poorer than would follow from the least favorable cases of the reduction of z with temperature. From the theory it follows that

$$\eta = \frac{T_h - T_c}{T_h} \cdot \frac{\sqrt{1 + \frac{1}{2}z(T_h - T_c)} - 1}{\sqrt{1 + \frac{1}{2}z(T_h + T_c)} + \frac{T_c}{T_h}},$$ (3)

i.e., the efficiency of thermoelectric generators should increase with increase of the thermodynamic coefficient. However, in many cases there is at some temperature a maximum of η. On further increase of the temperature difference $(T_h - T_c)$ between the hot (T_h) and cold (T_c) ends, the efficiency begins to fall. To explain this the following qualitative mechanism of the effect has been proposed: As a result of the quite strong temperature dependence of ρ for these materials, the upper (hot) parts of a thermoelectric generator have a considerably higher ρ than the lower (cold) parts. The consequence of this is a mismatch between the short-circuit currents of the cold and hot ends, $E_c/\kappa_c\rho_c \neq E_h/\kappa_h\rho_h$, i.e., part of the useful power is additionally dissipated inside the thermoelectric generator.

---

[*] This paper was presented at the First Conference on Thermoelectricity in December 1960.

A numerical calculation of the dependence of η on $(T_h - T_c)$ with allowance for the variation of ρ with temperature, in the case of a thermoelectric generator using $Bi_2Te_3 - Sb_2Te_3$ and $Bi_2Te_3 - Bi_2Se_3$, shows a sharp drop in the increase of η with increase of $(T_h - T_c)$, but fails to account for the inflection of the curve. In view of this it has been suggested that a change in the thermal conductivity strongly affects the electrical and thermal matching of the cold and hot parts of a thermoelement.

It is known that the lattice thermal conductivity and the thermal conductivity due to the majority carriers in the materials used in thermoelements decrease with increase of temperature, the former due to increase of the amplitude of the thermal vibrations of the atoms, and the latter in accordance with the Wiedemann – Franz law, in the case when σ decreases faster than T^{-1}. However, even in the presence of several per cent of carriers of the opposite sign there is an additional thermal conductivity due to the transport of energy by pairs, which rises very strongly with temperature and in some cases may dominate the total thermal conductivity of the material:

$$\kappa_{pair} = \frac{\sigma_n \sigma_p}{\sigma_n + \sigma_p} \cdot \frac{(\Delta E + 4kT)^2}{T}. \tag{4}$$

Such a rise of the thermal conductivity reduces η sharply for two reasons. Firstly, the mismatch between the quantities $\kappa\rho$ for the cold and hot parts increases; secondly, due to the greatly increased thermal conductivity of the hot parts, a large part of the temperature gradient is concentrated in the cold parts. Consequently large parts of the thermoelement branches are at a high temperature, which increases further the gap between $(\kappa\rho)_c$ and $(\kappa\rho)_h$. Calculations carried out on the basis of the data on the temperature dependence of z gave a dependence of η on ΔT with an inflection (maximum) at 200°C.

From the considerations given above it follows that a thermoelectric generator made of a material with uniform composition has a low efficiency for temperature drops of 300-400°C for the following reasons:

1) due to the rapid fall of z with increasing temperature the value of z which determines the generator efficiency is considerably lower than the value measured at room temperature;

2) due to the large difference between the values of $\kappa\rho$ for the hot and cold parts, additional power is dissipated in the thermoelectric generator itself which, by reducing the numerator in Eq. (1), reduces the value of η;

3) as a result of the rapid rise of κ at high temperatures, the temperature gradient becomes such that large parts of the generator branches are at a high temperature, which increases further the disproportion between the values of $\kappa\rho$ of the cold and hot parts and also increases the power which is wastefully absorbed in the generator branches.

The first factor can be reduced, and the second and third almost completely removed, by selecting optimum compositions for each range of temperatures.

The efficiency of thermoelectric generators may be raised considerably if the main parameters of the material (α, ρ, κ) vary in such a way along the generator branch that at each part of the branch the optimum value of z is obtained for the given temperature. A less effective, but still very powerful (and much simpler), method of increasing the generator efficiency is a discontinuous variation of the properties of the material by means of the multistage construction of thermoelectric generators. The optimum working conditions for a multistage generator reduce to the following requirements:

1) in each of the stages the material should have a value of $\bar{z}$ which is maximum for the given temperature range;

2) the value of $\kappa\rho$ should not have large discontinuities between the stages;

3) contacts between the stages should have minimum thermal resistance in the case of power taken separately from each stage and minimum ohmic resistance in the case of series connection of the stages.

Fundamentals of the Design of Multistage Thermoelectric Generators

The design of a multistage thermoelectric generator differs from that of a single-stage generator by the fact that in the former case, apart from the usual matching of cross sections, it is also necessary to match the lengths of the neighboring stages so that each stage works in a given range of temperatures:

$$\frac{l_{1,1}}{l_{1,2}} = \frac{\kappa_{1,1}\Delta T_{1,1}}{\kappa_{1,2}\Delta T_{1,2}}, \quad \frac{l_{2,1}}{l_{2,2}} = \frac{\kappa_{2,1}\Delta T_{2,1}}{\kappa_{2,2}\Delta T_{2,2}}, \text{etc.},$$ (5)

where $l_{1,1}$, $l_{1,2}$, $l_{1,3}$ are the lengths of the stages in the positive branch; $l_{2,1}$, $l_{2,2}$, $l_{2,3}$ are the lengths of the stages in the negative branch. If $\Delta T_{1,1} = \Delta T_{1,2}$, $\Delta T_{2,1} = \Delta T_{2,2}$, the ratio of the lengths is equal to the ratio of the thermal conductivities of the materials at a suitable average temperature.

The ratio of the cross sections of the branches of a multistage generator may be expressed as follows:

$$\frac{s_{\mathrm{I}}}{s_{\mathrm{II}}} = \frac{\sum\limits_{i=1}^{n} \kappa_{2i}\Delta T_{2i}}{\sum\limits_{i=1}^{n} \kappa_{1i}\Delta T_{1i}} \cdot \sqrt{\frac{\sum\limits_{i=1}^{n} \bar{\kappa}_{1i}\bar{\rho}_{1i}\Delta T_{1i}}{\sum\limits_{i=1}^{n} \bar{\kappa}_{2i}\bar{\rho}_{2i}\Delta T_{2i}}} \cdot$$ (6)

In the case of a two-stage thermoelectric generator we have

$$\frac{s_{\mathrm{I}}}{s_{\mathrm{II}}} = \frac{(\kappa_{2,1}\Delta T_{2,1} + \kappa_{2,2}\Delta T_{2,2})}{(\kappa_{1,1}\Delta T_{1,1} + \kappa_{1,1}\Delta T_{1,2})} \sqrt{\frac{\bar{\kappa}_{1,1}\bar{\rho}_{1,1}\Delta T_{1,1} + \bar{\kappa}_{1,2}\bar{\rho}_{1,2}\Delta T_{1,2}}{\bar{\kappa}_{2,1}\bar{\rho}_{2,1}\Delta T_{2,1} + \bar{\kappa}_{2,2}\bar{\rho}_{2,2}\Delta T_{2,2}}} \cdot$$ (7)

The formulas (5), (6) and (7) are all that is necessary for the calculation of the lengths and cross sections of the branches in a two-stage generator.

In a more accurate expression for the ratio of l_{1i} and l_{2i} one should allow for the fact that at the stage contacts there is an emission and absorption of Peltier heat, proportional to πI, where π is taken at the temperature of the contact. To determine the true ΔT for each of the stages it is necessary to derive and solve the equation for the total heat balance for both (upper and lower) stages allowing for the temperature dependence of α, ρ, κ, and τ [2]. In the case when the average integrated values of these quantities are different for $\Delta T_{1,1}$ and $\Delta T_{1,2}$, the Joule and Thomson heats in each of the stages are not strictly shared equally.

Operation of Thermoelectric Generators in Various Temperature Intervals

We have determined the optimum carrier densities for materials intended for operation in various temperature intervals on the basis of the calculated temperature dependences of z.

In selecting the parameters for the individual stages, our attention was directed principally to the condition that the value of z should be a maximum for the given range of temperatures. Therefore the quantity $\kappa\rho/\Delta T$ had as a rule discontinuity at the stage contacts, but the values of $\kappa\rho$ for the cold and hot junctions in a two-stage thermoelectric generator differed much less than in the case of a single stage subjected to the same total temperature drop.

Thus, for example, $(\kappa\rho)_{300°K} = 3.27$ (in arbitrary units), $(\kappa_1\rho_1)_{450°K} = 6.15$, $(\kappa_1\rho_1)_{600°K} = 12.6$. For the "hot" stage $(\kappa_2\rho_2)_{300°K} = 2.18$; $(\kappa_2\rho_2)_{450°K} = 4.1$; $(\kappa_2\rho_2)_{600°K} = 7.5$.

When the lower stage is subjected to the total temperature drop $\Delta T = 300°$, we have $(\kappa_1\rho_1)_{600°K}/(\kappa_1\rho_1)_{300°K} = 3.85$; for the upper stage under $\Delta T = 300°$, we have $(\kappa_2\rho_2)_{600°K}/(\kappa_2\rho_2)_{300°K} = 3.45$.

When the stages are connected in series (i.e., $\Delta T = 150°$ at each stage), then $(\kappa_2\rho_2)_{600°K}/(\kappa_2\rho_2)_{300°K} = 2.3$, i.e., we obtain a more favorable ratio.

For the negative branch we obtain an even better change in properties: $(\kappa_2 \rho_2)_{600°K} / (\kappa_2 \rho_2)_{300°K}$ = 1.4 instead of 3.1 for the lower stage and 2.95 for the upper stage when each of them is subjected to the full 300°.

The integral efficiency was determined for generator pairs made of a material with n ranging from $1.5 \cdot 10^{19} \, cm^{-3}$ to $1.4 \cdot 10^{20} \, cm^{-3}$ in order to select the optimum compositions for various temperature drops.

The most typical dependences $\eta (\Delta T)$ are given in Fig. 1. Curve 4 shows the dependence of η for a thermoelement made of materials with n = $1.8 \cdot 10^{19} \, cm^{-3}$, $\alpha_p = 218 \, \mu V/deg$, $\sigma_p = 600$ ohm$^{-1} \cdot$ cm^{-1}, $\alpha_n = 215 \, \mu V/deg$, $\sigma_n = 520$ ohm$^{-1} \cdot$ cm^{-1}.

Such an element gives quite a high efficiency for a temperature drop of 100° (300-400°K) but because of the early onset of intrinsic conduction the value of η ceases to increase with ΔT even at $\Delta T = 200°$, and beyond this temperature drop the efficiency begins to fall because of the strong mismatch of $\kappa\rho$ values of the hot and cold ends. Curve 3 represents n = $3.2 \cdot 10^{19}$ cm^{-3}, $\alpha_p = 195 \, \mu V/deg$, $\sigma_p = 1300$ ohm$^{-1} \cdot$ cm^{-1}, $\alpha_n = 165 \, \mu V/deg$, $\sigma_n = 1800$ ohm$^{-1} \cdot$ cm^{-1}. This thermoelement with a slightly lower efficiency for $\Delta T = 100°$ gives in practice the same total efficiency as the thermoelement with lower n because of the later onset of the undesirable effects related to intrinsic conduction. Curve 1 represents n = $1 \cdot 10^{20}$ cm^{-3} ($\alpha_p = 105 \, \mu V/deg$, $\sigma_p = 2590$ ohm$^{-1} \cdot$ cm^{-1}, $\alpha_n = 130 \, \mu V/deg$, $\sigma_n = 1500$ ohm$^{-1} \cdot$ cm^{-1}); for $\Delta T = 100°$ (400-500°K) the values of η are smaller but the curve shows no tendency to "saturation." Curve 2 gives the temperature dependence of η for $\Delta T = 400°$ for materials with n $\approx 4 \cdot 10^{19}$ cm^{-3}.

Figure 2 gives the dependences of the integral and differential efficiencies for a thermoelement. The differential efficiencies for $\Delta T = 100°$ (300-400°K, 400-500°K, and 500-600°K) are plotted as continuations of one another. In spite of the very low efficiencies at high temperatures (since the composition at these temperatures is unsuitable), the total efficiency is almost 1.5 times greater than η_{max} for a single stage over the same range of temperatures.

Figure 2 shows the first attempt to replace a single stage with a two-stage generator. Examination of the curves shows that for the first 150° of the temperature drop it is more advantageous to use the material which has been selected for relatively low temperatures, but beginning from 200°C the advantage of the two-stage generator is beyond doubt.

In assembling multistage generators, their cold junctions are plated with an alloy Bi–Sn, the contacts between stages with Sn, and the hot junction with Bi. Later it was found experimentally that it is more convenient to wet all the junctions and contacts with an In – Ga alloy having a melting point of $\approx$ +16°C. This alloy has a very low diffusion coefficient, low vapor pressure, and wets well the materials used in thermoelements [3].

To solve the problem of the magnitude of the efficiency obtained from two electrically isolated stages, compared with a thermoelectric generator assembled from the same two stages connected electrically and thermally in series, we carried out calculations and tests for two stages, each at $\Delta T = 150°$ (300-450 and 450-600°K), and then assembled them into a two–stage generator. The results are shown in Fig. 3. The efficiencies in the two temperature ranges are superimposed on one another and the sum gives $\eta = 8.1\%$ for $\Delta T = 300°$. The lower (cold) stage, having the properties $\alpha_p = 230 \, \mu V/deg$, $\sigma_p = 750$ ohm$^{-1} \cdot$ cm^{-1}, $\alpha_n = 200 \, \mu V/deg$, and $\sigma_n = 800$ ohm$^{-1} \cdot$ cm^{-1}, gave $\eta = 5.3\%$ for $\Delta T = 150°$. The upper stage ($\alpha_p = 140 \, \mu V/deg$, σ_p = 2200 ohm$^{-1} \cdot$ cm^{-1}, $\alpha_n = 145 \, \mu V/deg$, $\sigma_n = 1100$ ohm$^{-1} \cdot$ cm^{-1}) gave $\eta = 2.8\%$ for $\Delta T = 150°$. A two-stage generator consisting of the same two stages gave a value of $\eta = 7.8\%$ for the same total temperature drop, i.e., only 0.3% lower than the two stages separately (in Fig. 3 the point denoted by a triangle represents the two-stage generator).

It should be noted that the efficiency of the separate stages (working individually) [7] is expressed as

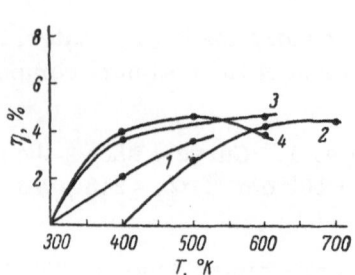

Fig. 1. Dependence of the efficiency of various thermoelements on the hot-junction temperature (for $T_c = 300°K$).

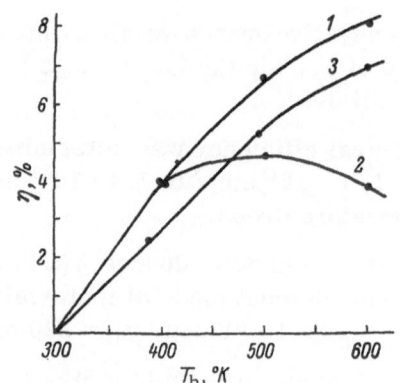

Fig. 2. Differential (1) and integral (2, 3) efficiencies of single-stage (2) and two-stage (3) thermoelements.

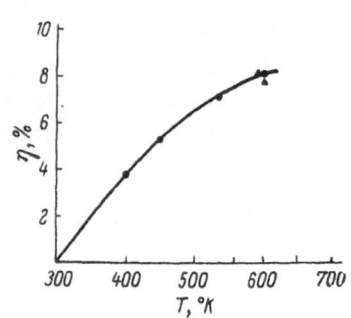

Fig. 3. Dependence of the efficiency of a two-stage thermoelement on the hot-junction temperature (for $T_c = 300°K$).

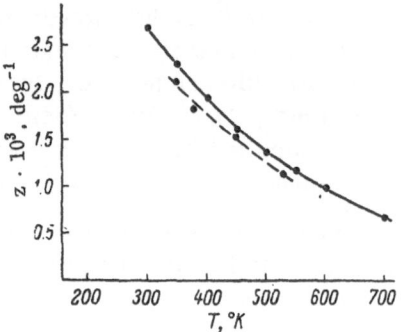

Fig. 4. Comparison of the experimental and calculated temperature dependences of $z_{optimum}$.

$$\eta = \eta_{i1} + \eta_{i2} - \eta_{i1}\eta_{i2}, \tag{8}$$

and consequently the upper stage reduces the total heat flowing to the lower stage by $(1 - \eta_1)$.

In our case $\eta_1\eta_2$ amounts to $\approx 0.15\%$ (which follows from the 0.3% difference), i.e., multistaging had practically no effect on η_{tot} of the generator. However, it should be noted that this is the consequence of a fortunate combination of $\kappa_1\rho_1$ and $\kappa_2\rho_2$ as well as of a relatively small temperature drop. In general (particularly in the case of high ΔT), multistaging with separate power take-off from each stage should improve the efficiency considerably compared with the series connection of the stages.

A somewhat better result was obtained for a two-stage thermoelectric generator in which the stages of the positive branch were working under the same temperature drop of $\Delta T = 150°$ and the negative-branch stages were subjected to $\Delta T = 100°$ (lower stage) and $\Delta T = 200°$ (upper stage). This thermoelement gave $\eta = 8.1\%$ (the black triangle in Fig. 3) for an average temperature drop of $270°$ (35-305°C).

From the efficiencies of the thermoelements we can determine $\bar{z}$ in the working range of temperatures.

Transforming Eq. (1) we have
$$z = \frac{2\eta(2 + \eta)}{(T_h - T_c)\left[1 - \eta\dfrac{T_h}{T_h - T_c}\right]}. \tag{9}$$

According to Eq. (9), when $\eta = 4\%$ for $\Delta T = 100°$ (300-400°K), $\bar{z} = 2.17 \cdot 10^{-3} \text{deg}^{-1}$; when $\eta = 5.3\%$ for $\Delta T = 150°$ (300-450°K), $\bar{z} = 1.8 \cdot 10^{-3} \text{deg}^{-1}$; when $\eta = 2.8\%$ for $\Delta T = 150°$ (450-600°K), $\bar{z} = 1.15 \cdot 10^{-3} \text{deg}^{-1}$; when $\eta = 8.1\%$ for $\Delta T = 270°$ (310-580°K), $\bar{z} = 1.53 \cdot 10^{-3} \text{deg}^{-1}$.

Figure 4 shows the experimental values (dashed curve) in comparison with the calculated temperature dependences for each range of temperatures.

In our experiments the efficiency of thermoelectric generators was raised by a more rational selection of the parameters of the materials and the use of multistage generators.

The multistage generator branches having the same composition as host material differed only in the concentration of alloying impurity. To obtain a further improvement of the efficiency and to extend the range of working temperatures of multistage thermoelectric generators, it is necessary to assemble generators with three or more stages. This requires the use of new materials efficient at higher temperatures, with the necessary prerequisite a study of the temperature dependences of the principal thermoelectric properties of such materials.

LITERATURE CITED

1. A. F. Ioffe, Semiconductor Thermoelements, Izd. Akad. Nauk SSSR, Moscow-Leningrad, 1956; English translation: Infosearch Ltd., London, 1957.
2. B. Ya. Moizhes, Fiz. Tverd. Tela 2:728, 1960.
3. E. K. Iordanishvili, Author's Certificate No. 660136, 1960.

GALVANIC METHOD OF THERMOELEMENT BRIDGING *

A. D. Finogenov

Substances used currently for the manufacture of the positive and negative branches (arms) of thermoelements are, as a rule, complex chemical compounds of nonstoichiometric composition, containing various impurities which strongly affect their thermoelectric properties [1].

The thermoelectric properties of finished thermoelement branches are often highly sensitive to heat treatment, especially in air, due to the possibility of oxidation.

Therefore it follows that the alternate connection ("bridging") of the positive (2 in Fig.1) and negative (3) branches of a thermoelement by means of current-carrying bridges (1 and 4) should be carried out in such a way as to avoid heat treatment.

Heat treatment is unavoidable in the case of bridging by soldering, the pressing together of two parts, or the application of a melt. In soldering, the end of a semiconductor sample is heated to the melting temperature of the solder. The pressing together of two parts is usually carried out in heated molds in order that the plastic deformation of the semiconductor and the bridge material occur at relatively low pressure. When applying a melt of the bridging material to the ends of the thermoelement branches (in particular, a metal melt) for the purpose of forming a current-carrying bridge, the semiconducting material softens at the point of contact with the metal if the melting point of the semiconductor is lower than that of the metal. In some cases, this may produce a considerable deterioration of the thermoelectric properties of the thermoelement by altering the composition near the metal — semiconductor boundary.

The electrolytic deposition of metal bridges on semiconductors allows us to overcome all these difficulties since in depositing such metals as Ni, Fe, Cu, Zn, etc., the thermoelement branches being processed have the same temperature as the electrolyte in which they are immersed, i.e., 20-90°C, at which temperature there should be no changes of the thermoelectric properties of the semiconducting materials involved. Moreover, the electrical conductivity of metals deposited electrolytically is, as a rule, higher than that of cast or pressed metals [2]. This is important because a principal factor determining the quality of a thermoelement is the resistance of the current-carrying bridges; it must be minimal as it augments the internal resistance of the thermoelement, which consists mainly of the resistance of the positive and negative branches. Furthermore, the galvanic method can be used to join the branches of thermoelements selected most rationally for a given substance so that its thermoelectric properties are fully exploited.

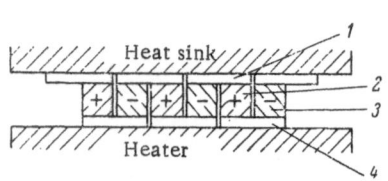

Fig. 1. Series connection of thermo-element branches. 1 and 4 are the current bridges; 2 is the p-type thermo-element branch; 3 is the n-type branch.

The branches may be pressed, cast, or single crystal; they may have very different melting points; and they may have been subjected to different heat treatments.

* This paper was presented at the First Conference on Thermoelectricity in December 1960.

Using the electrolytic method for bridging, we can employ the standard electroplating [2-5] and electroforming [6] methods and electrolytes, with due allowance for the special properties of the semiconducting materials.

Thus, for example, substances such as SbZn, $CoSb_3$, Sb, Bi, may be degreased electrochemically. On the other hand, the side reactions occurring at the cathode forbid the use of electrochemical degreasing in the case of such substances as Te, Se, Sb_2Te_3, Bi_2Te_3, PbSe, PbTe, the ternary alloy $Bi_2Te_3 - Sb_2Te_3$ used for positive branches [7, 8], and the ternary alloy $Bi_2Te_3 - Bi_2Se_3$ used for negative branches [9]. In these cases we may use different degreasing methods: treatment with French chalk, organic solvents, chemical degreasing by boiling in an aqueous solution of soda with dextrin, etc. To etch all these materials, it is necessary to select suitable compositions because the etchants recommended in the literature for nonferrous metals [2, 4] are usually unsuitable.

In depositing metals on semiconducting compounds, we meet with difficulties due to the fact that frequently the same electrolyte has to be used to establish a metal bridge between thermoelement branches which differ very strongly in their electrode potential in a given electrolyte, although it is known [2, 10] that the electrode potential is related to such effects as cementation (contact deposition) and pitting.

The phenomenon of cementation consists of the electrode material of the cathode being dissolved because of its chemical activity, displacing the electrolyte cations, which are then deposited on the cathode, even in the absence of a current, forming a loose layer weakly bound to the cathode.

The bridging is adversely affected also by the form of pitting connected with the evolution of hydrogen ions at the cathode, which, remaining for a long time at the cathode, may give rise to the formation of continuous (through) channels in the bridge, and thereby may leave parts of the ends of the thermoelement branches unprotected by a metal layer. This may have a deleterious effect in the case when a layer of one metal is deposited over another and the former reacts easily with the semiconductor. The methods of preventing pitting and cementation are dealt with satisfactorily in the literature [2, 4].

Thermoelements prepared from the semiconducting materials mentioned above can be connected by means of nickel bridges formed in a bath of 400 g $NiSO_4 \cdot 7H_2O$, 15 g $NiCl_2 \cdot 6H_2O$, 30 g H_3BO_3, 0.03 g $CdSO_4$, 1000 ml H_2O, at t = 50-60°C, a current density of 10-15 A/dm^2 (with constant mixing), a pH of 3.5-4, and with nickel anodes.

There are several variants of the method of formation of current-carrying bridges by the galvanic method.

1. Deposition of a Metal from Electrolyte Directly on the Ends of Thermoelement Branches

Single-Layer Metal Bridge. In this case, the positive (1) and the negative (3) branches of a thermoelement (Fig. 2), separated by an insulating spacer 2, are first joined by an auxiliary current-carrying bridge 4. This bridge is used to establish electrical contact between the thermoelement's branches in the process of covering their ends with a metal 5 in an electrolytic bath.

To produce an auxiliary current-carrying bridge we may use the methods of coating with graphite, sputtering in vacuum, chemical deposition of silver, nickel, and other metal films, soldering, etc. [6].

Thermoelements based on alloys consisting of 74 mol.% Sb_2Te_3 and 26 mol.% Bi_2Te_3 with an excess of 3 wt.% Te (p-type) or Bi_2Te_3 (n-type) with a nickel current-carrying bridge 5 (Fig. 2) operate with satisfactory stability for a long time at a temperature of 250°C at the hot end of the thermoelement and 20°C at the cold end. The resistance of thermoelements 8 mm high and

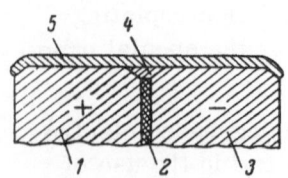

Fig. 2. Single-layer metal bridge: 1 is the p-type branch; 2 is an insulating layer; 3 is the n-type branch; 4 is an auxiliary current bridge; 5 is the metal bridge.

0.64 cm^2 in cross section rose by 10% during a test lasting 5100 hr, compared with the initial resistance; this was mainly due to the diffusion in the nickel — semiconductor contact layer. During the same period the thermoelectric power of the branches rose by 6-8%. (The thermoelement branches were prepared by the powder metallurgy method.)

At higher temperatures it is not advisable to use nickel in contact with antimony and bismuth tellurides and selenides because the process of diffusion is more intense than at 250°C and the total resistance of the thermoelements consequently rises faster due to the change in the resistance of the contact layers.

Such substances as SbZn (p-type), $CoSb_3$, and PbTe (n-type) can withstand higher temperatures in contact with nickel than can antimony and bismuth tellurides and selenides. Thus, in the case of 300°C at the hot end of the thermoelement and 20°C at the cold end, the resistance of a branch made of SbZn did not change in 3000 hr, the resistance of a branch made of $CoSb_3$ rose by 0.7% in 3500 hr, and the resistance of a branch made of PbTe rose by 8-10% in 4600 hr, this being accompanied by an increase in the thermoelectric power by 7-8%.

SbZn, $CoSb_3$, and PbTe branches were prepared by the powder metallurgy method and were 8 mm high and 0.64 cm^2 in cross section.

Two-Layer Metal Bridge. For the purpose of bridging, it is desirable to use a material with the highest possible electrical and thermal conductivities. The most suitable material is copper.

However, copper deposited by the galvanic method on samples of bismuth and antimony tellurides and selenides diffuses into these compounds even at room temperature, impairing considerably their thermoelectric properties in a very short time, varying from several hours to several days, depending on the composition and structure of the substance in contact with the copper. In fact, Hansen [11] showed that in Bi_2Te_3 the diffusion coefficient of copper along a direction perpendicular to the cleavage planes is $D_\perp = 3 \cdot 10^{-15} \text{ cm}^2 \cdot \text{sec}^{-1}$, while along a direction parallel to these planes it is considerably greater: $D_\parallel = 10^{-6} \text{ cm}^2 \cdot \text{sec}^{-1}$. Therefore, to protect a semiconducting material from the action of copper on the thermoelement ends, it is necessary to deposit a layer of metal which does not interact with the given substance in the working range of temperatures (for example, nickel) and then to deposit a layer of copper. In this case, the working life of the thermoelement is governed by the rate of diffusion of copper into the semiconductor through the nickel layer.

2. Deposition of a Metal in an Electrolytic Bath on a Protective Bridge

Here, the protective bridge in direct contact with the semiconductor branches may be prepared from some material by pouring out a melt, pressing two parts together, etc. On top of this bridge, we deposit electrolytically a layer of a metal with a higher electrical conductivity, and consequently a higher thermal conductivity than the material of the protective bridge.

This makes it possible to reduce considerably the total thickness of the current-carrying bridge, the consumption of an expensive material which may be necessary to establish a stable contact with the semiconductor, and the temperature drop across the current bridge.

Thus, for example, when bridging was achieved by means of molten metal it was found that a thermoelement made from the compounds $Bi_2Te_3 - Sb_2Te_3$ and Bi_2Te_3 with bridges consisting of the 98 wt.% Sb +2% Pb alloy were quite stable up to 300°C; a small amount of lead — within the limits of solubility — was introduced in an attempt to reduce the melting point of the

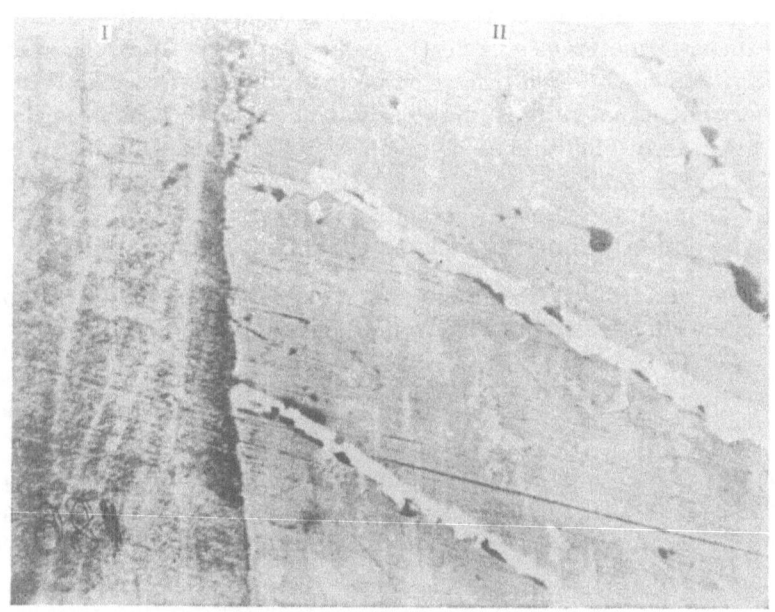

Fig. 3. Contact of the $Bi_2Te_3 - Sb_2Te_3$ alloy with electrolytically deposited antimony. I) Sb; II) $Bi_2Te_3 - Sb_2Te_3$ alloy. The sections where etched with a solution consisting of equal parts by volume of H_2O and HNO_3. 136x.

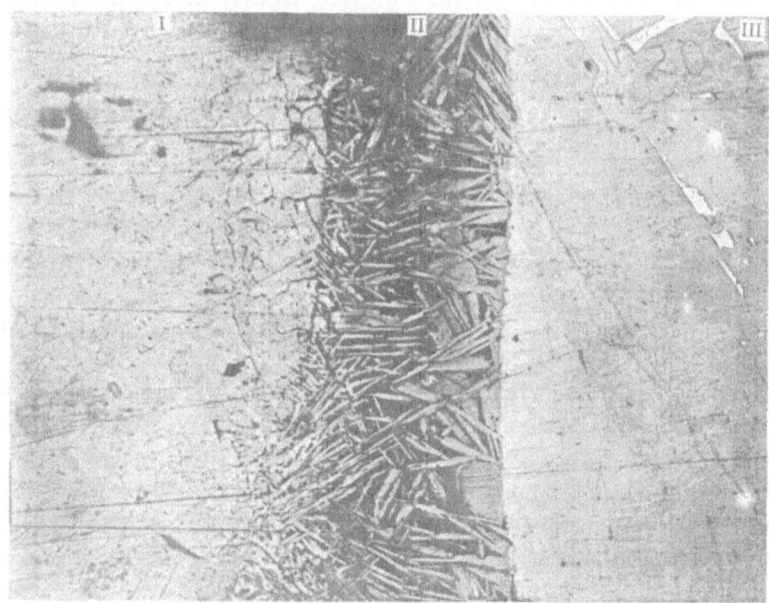

Fig. 4. Contact of the $Bi_2Te_3 - Sb_2Te_3$ alloy with antimony deposited in the molten state. I) Sb; II) transition melted layer consisting approximately of 50 wt.% Sb + 50 wt.% $BiTe_3 - Sb_2Te_3$; III) $Bi_2Te_3 - Sb_2Te_3$ alloy. Etching was carried out in the same way as for Fig. 3. 136x.

bridging alloy [12] so that the structure and composition of the intermediate molten layer is different from that of pure antimony and its thickness is smaller. The power of thermoelements with branches of 1 cm^2 cross section (the branches were prepared by directional crystallization in molds), 6 mm high and 3 mm thick bridges fell by 15% during 2100 hr with 300°C at the hot end and 70°C at the cold end. The power reduction occurred due to oxidation, partly by deterioration of the parameters of the branches and partly due to the increase of the resistance of the intermediate molten layer between the antimony and semiconductor.

Thus the 98% Sb + 2% Pb alloy is sufficiently stable under operating conditions, but its main disadvantage is its low electrical conductivity: $3 \cdot 10^4$ ohm$^{-1} \cdot$ cm^{-1}. On the other hand, the electrical conductivity of nickel is 5 times higher and its contact with antimony is stable for 5000 hr at 400°C. Hence, it is possible to make a thermoelement with an antimony bridge 0.8-1 mm thick with a 0.4-0.5 mm thick nickel layer on top of it; the latter is equivalent to a layer of antimony 2 mm thick. The service life of such a thermoelement is increased by the application of the nickel layer since in the use of a thinner commutation bridge the thickness of the melted layer between the commutation alloy and the semiconductor is reduced.

In summarizing the results reported above on the galvanic method of deposition of a current bridge, two important disadvantages of this method should be noted:

1. The galvanic method is convenient for the deposition of metals such as Ni, Cu, Fe, Zn, etc., but the deposition of alloys (brass, bronze, etc.) is more difficult and the difficulties are even greater in the case of more complex compounds.

2. The low bonding forces at the metal – semiconductor contact. To increase the extent of the bond it is necessary to bevel slightly the side surfaces of the thermoelement branches next to the ends. This is an important limitation of the method in the case of thermoelements of low height.

The advantages of the galvanic method of bridging thermoelements are as follows:

1) the presence of a sharp metal – semiconductor boundary (Fig. 3), and the absence of intermediate layers at this boundary; for comparison, Fig. 4 shows the metal – semiconductor boundary when the current bridge was prepared by applying a molten layer in air;

2) the high electrical conductivity of the electrolytically deposited metal layers;

3) the possibility of deposition at 20-90°C of metals with relatively high melting points.

In conclusion, the author thanks A. N. Voronin for his interest in this work and discussion of the results, and I. M. Krasheninnikova, who carried out the microstructure analysis in the life tests of the thermoelements.

LITERATURE CITED

1. A. F. Ioffe, Semiconductor Thermoelements, Izd. Akad. Nauk SSSR, Moscow-Leningrad, 1960.
2. V. I. Lainer and N. T. Kudryavtsev, Fundamentals of Electroplating, Metallurgizdat, 1953.
3. Yu. V. Baimakov, Electrolysis in Metallurgy, Metallurgizdat, 1939, Vol. I.
4. K. S. Goncharenko, Short Handbook for Electrochemical Workers, Mashgiz, Kiev-Moscow, 1955.
5. G. T. Bakhvalov et al., Handbook for Electrochemical Workers, Metallurgizdat, 1954.
6. B. Ya. Kaznachei, Electroforming in Industry, Rosgizmestprom, 1955.
7. G. I. Shmelev, Alloys Based on Ternary Compounds for Thermoelements, Dissertation for Candidate's Degree, Moscow, 1949.
8. G. V. Kokosh and S. S. Sinani, Fiz. Tverd. Tela 1:89, 1959.
9. G. N. Gordyakova, G. V. Kokosh, and S. S. Sinani, Zhur. Tekh. Fiz. 28:3, 1958.
10. Berl'-Lunge, Technochemical Investigation Methods, Ob'edinenie Nauch.-Tekh. Khimteoret, Leningrad, 1937.
11. R. O. Carlson, J. Phys. Chem. Solids 13:65, 1960.
12. M. Hansen, Constitution of Binary Alloys [Russian translation], Metallurgizdat, 1941, Vol. II.

NEW THERMOELECTRIC COOLING DEVICES

E. A. Kolenko

Thermoelectric cooling is undoubtedly one of the youngest branches of modern technology. The first engineering designs of thermoelectric cooling devices were developed in 1956 at the Institute for Semiconductors of the USSR Academy of Sciences. In the last five years, over 60 various cooling devices have been developed, some of which are already being manufactured on an industrial scale. The devices developed so far are intended for use in nuclear physics, astronomy, vacuum technology, metallurgy, metrology, medicine, botany, and many other branches of science and technology. Currently, simultaneously with the development of new cooling devices, intensive work is being done on the improvement of the basic parameters of materials used to make thermopiles, and on the development of rational methods of device construction and assembly. Thus, for example, in 1961 the efficiency of thermoelectric materials was considerably improved, making it possible to obtain a temperature drop of 50-55° by means of a thermocouple. Methods for heat removal — electrically insulated thermal contacts with low thermal resistance — have been developed, which improved the basic technical thermal parameters of the devices. The introduction of special damping layers of lead prevented mechanical stresses in thermoelements, which had earlier resulted in fracture. The development of special bridging solders based on antimony and bismuth made it possible to simplify considerably one of the main operations in the manufacture of thermoelectric cooling devices, i. e., the bridging of a thermoelectric battery. New materials — polyurethane and foamed polystyrene — have been used as heat insulation materials.

These advances in the technology of thermoelectric cooling devices were fundamental to the devices developed in 1961, some of which are briefly described below.

A Cooling Device for the Determination of the Freezing Point of Petroleum Products

One of the main parameters affecting the quality of performance of petroleum products, in particular diesel fuel, is the freezing temperature. The currently used industrial method for measuring the freezing temperature is based on the determination of the onset of attenuation of an ultrasonic pulse produced in a special cooled cuvette filled with the product to be investigated. The cuvette temperature should be reduced gradually, and after reaching the freezing point the petroleum product should be heated for several minutes and removed from the cuvette. The cuvette is then filled with a new batch of the product and the measurements are repeated. It should be added that the whole process of cooling and heating of the cuvette, filling and decanting the petroleum product, and the determination of the freezing temperature should be carried out automatically.

In cooperation with the Kuibyshev Scientific Research Institute for the Petrolum Industry, the Institute for Semiconductors of the USSR Academy of Sciences has developed a thermoelectric device which satisfies the requirements stated above.

Two two-stage thermopiles are soldered to the two opposite sides of a cuvette which has connecting branches for filling and decanting the petroleum product, apertures for an ultrasonic transducer, and a microthermistor for measuring the temperature of the product in the

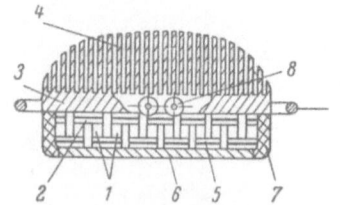

Fig. 1. A section through a microcooler for the treatment of skin
diseases.

cuvette. To increase the cooling efficiency of the second stages
of the piles, the first and second stages are supplied in series.
This is done by means of special electrically insulated bridging
plates which have low thermal resistance. These bridging plates
make it possible to solder directly the thermopiles to the cuvette
and the heat removal system, which reduces to zero the stray
thermal resistances and makes it possible to manufacture a
single mechanically strong unit. Heat is removed from the hot
junctions of the thermoelements in the first stages by means of
running water, the passages for which are located directly in
the hot bridging plates.

The basic parameters of this cooling device for the determination of the freezing point
of petroleum products are as follows: 1) working current under cooling conditions: 40 A; 2)
voltage drop under cooling conditions: 1.66 V; 3) power consumed: 68.4 W; 4) minimum temperature in the working chamber (when the water temperature is +20°) − 34°; 5) minimum temperature drop produce by the device: 54°; 6) time to establish the minimum temperature: 35
min; 7) working current to establish a temperature of +10° in the chamber: 15 A; 8) time to establish +10°: 6 min; 9) water consumption: 100 liters/hr; 10) volume of the working chamber:
33 cm^3; 11) dimensions: 117 × 100 × 110 mm; 12) weight: 1560 g.

Microcooler for the Treatment of Skin Diseases

Some skin diseases may be successfully treated by local cooling. If the temperature of
an area of the skin is lowered by 8-10° with respect of the body temperature, the consumption
of nutrients by the affected area is reduced and healing starts. Depending on the nature of the
disease, the cooling should be applied for a period from several weeks to several months. Obviously, a cooling device suitable for this purpose should be small and light in order not to
handicap the patient.

For this purpose, a self-contained thermoelectric device has been developed, a section
through which is shown in Fig. 1. The main part of the device is a single-stage thermopile
1, consisting of 12 thermoelements joined in series by means of electrically insulated plates 2.
The hot junctions of the thermoelements are soldered to an aluminum radiator system 3 fitted
with a number of fins 4. The cold junctions of the thermoelement are soldered via electrically
insulated bridging plates 5 to a collector 6 which forms the working part of the device. The
device is made mechanically strong by a decorative plastic ring 7. The power supply is obtained from a small silver − zinc storage cell connected to terminals 8.

The device is attached to the affected area of the skin by means of a special strap. Figure
2 shows the general view of the cooler for the treatment of skin diseases.

The basic parameters of the device are as follows: 1) working current: 3 A; 2) working
voltage: 0.4 V; 3) power consumed: 1.2 W; 4) working surface: circle of 50 mm diameter; 5)
dimensions: 80 mm in diameter, 47 mm high; 6) weight:
360 g; 7) weight including the storage cell: 1180 g.

Fig. 2. General view of a microcooler for the
treatment of skin diseases.

Microcooler for Laboratory Use

In laboratory work, it is frequently necessary
to carry out a study in a certain range of temperatures.
The existing heat chambers are large, consume considerable power, and, most important, cannot guaran-
tee continuous temperature control over the whole
working range. Consequently, a small thermoelectric
cooler was developed which makes it possible to vary
temperature, with any required accuracy and according
to any program, between −40° and +50°.

The thermopile of the device consists of two stages connected in series with each other by means of electrically insulated bridging plates. The working chamber of the cooler, a copper cylinder, is soldered directly to the cold junctions of the second stage of the thermopile. The heat is removed from the hot junctions of the pile by running water through the aluminum base of the device to which the bridging plates of the hot junctions of the battery are soldered. To ensure the necessary heat pumping rate, the first stage of the battery contains 18 thermoelements, and the second 3 thermoelements. The working volume of the device is finished outside by a layer of thermal insulation consisting of foamed plastic and "mipor" (microporous rubber). A demountable thermally insulated cover gives access to the working chamber of the device.

The basic parameters of the device are: 1) working chamber volume: 75 cm³; 2) maximum working current: 50 A; 3) voltage drop across the device: 1.4 V; 4) power consumed: 70 W; 5) minimum temperature in the working chamber (when the cooling-water temperature is +15°):−40°C; 6) maximum temperature in the working chamber: +50°C; 7) working current under maximum heating conditions: 8 A; 8) water consumption: 30 liters/hr; 9) dimensions (diameter × height): 110 × 120 mm; 10) weight: 1.5 kg.

The general view of the microcooler for laboratory use is given in Fig. 3.

Thermoelectric Traps for Diffusion Pumps

To freeze out the residual working oil vapors in oil-vapor diffusion pumps, it is usual to employ traps cooled with liquid nitrogen. However, it is known that sufficiently complete condensation of the residual oil vapors is obtained even at temperatures of −40 to −42°. In view of this, the Institute for Semiconductors of the USSR Academy of Sciences has, since 1959, developed high-vacuum thermoelectrically cooled traps. Traps have been developed for the following pumps: MM-40A, TsVL-100, N-5, N-2T, N-5T, N-8T.

The traps for the pumps MM-40A, TsVL-100, and N-5 have been in continuous production for over two years. In 1961-2, the development of thermoelectric traps for the powerful pumps N-2T, N-5T, and N-8T was completed.

These traps, which differ from one another in their dimensions, power supplies, and some constructional details, consist of two-stage batteries with parallel supply of the stages. The number of thermoelements in the first and second stages is governed by the power supply conditions and the required heat-pumping rate. To reduce the resistance of the trap to the flow in the pump, the former is made in the form of a "barrel," which increases considerably the feed-through cross section of the trap. The heat is removed from the hot junctions of the thermopile by running water taken from the cooling system of the pump. The heat removal system consists of copper double-U-shaped boxes soldered vacuum-tight to the trap casing. Water passes through the inner channels of these boxes.

The trap for the N-2T pump has two such boxes on which a two-stage pile is mounted.

The traps for the pumps N-5T and N-8T have three boxes and three-stage piles. Special bridges join in series the water channels in the heat-removal boxes.

The bridging plates of the thermoelements in the first and second stages of batteries have copper con-

Fig. 3. General view of a microcooler for laboratory use.

Basic Rating Data of Thermoelectric Vacuum Traps

Type of trap	Working current, A	Voltage drop, V	Temperature, °C		Resistant to gas flow, %	Water consumption liters/hr	Weight, kg	Dimensions	
			stage I	stage II				height	diameter
TVL-2T-1 . .	120	3.6	−20	−50	45	150	35	210	490
TVL-5T-1 . .	120	6.6	−18	−50	45	150	120	330	765
TVL-8T-1 . .	150	8.3	−19	−49	45	150	240	390	1000

Note. Values of the temperatures in the stages I and II were obtained for the cooling-water temperature of +18°C and the pressure P = 10^{-5} mm Hg.

densation surfaces. The condensation surfaces of the first and second stages make an angle of 45° with the vertical. Such a system of condensation surfaces has a low resistance to the gas flow but makes the trap "impermeable" to direct penetration of the oil vapor molecules from the pump to the chamber being evacuated. In the diffusion pumps N-2T, N-5T, and N-8T, the residual oil vapor represents a fairly large thermal load on the thermopile, reducing the efficiency of the latter. Consequently, in these traps, an additional system of condensation surfaces is fitted directly to the heat-removal boxes. Thus the main heat load is taken up by the first system of the condensation surfaces which are at the temperature of the running water and therefore have a high heat-pumping rate. To reduce to minimum the stray heat losses between the thermopile and the heat-removal system, the first stage of the pile is soldered directly to the heat-removal boxes via electrically insulated thermal contacts having low thermal resistance even at considerable rates of heat flow.

The basic parameters of the thermoelectric high-vacuum traps for powerful diffusion pumps are listed in the table.

METHOD OF BRIQUETTING AND SUBSEQUENT HEAT TREATMENT OF THERMOELEMENT BRANCHES MADE OF $Bi_2Te_3 - Sb_2Te_3$ AND $Bi_2Te_3 - Bi_2Se_3$ ALLOYS

A. N. Voronin and R. Z. Grinberg

Three methods are currently used to manufacture thermoelement branches from alloys based on tellurides. The first method, which is most widely used, is powder metallurgy. The alloy is ground to grain dimensions of less than 2 mm, poured into a mold heated to 400-430°C, and pressed for 5 min at 4-7 tons/cm². The second method of manufacturing thermoelement branches is casting under conditions of directional crystallization. The third method is the preparation of single-crystal or nearly single-crystal thermoelement branches. As is known, single-crystal thermoelements have the best thermoelectric properties and the highest values of z compared with thermoelements prepared by other methods.

The Special Design Office of the Institute for Semiconductors of the USSR Academy of Sciences has developed a new variant of the powder metallurgy method for $Bi_2Te_3 - Bi_2Se_3$ and $Bi_2Te_3 - Sb_2Te_3$ alloys: cold pressing of thermoelement branches followed by heat treatment. The ground alloy is poured into a mold at room temperature and a definite pressure is applied to it. The duration of application of the pressure does not greatly affect the electrical properties of the sample. The duration actually used was 10 sec. Next the briquets were placed in an evacuated container (at 10^{-1}-10^{-2} mm Hg) where they were fired and annealed for a certain time. Cooling took place in the furnace.

It was necessary to determine the relationship between the grain size, pressure, temperature of heat treatment, duration of heat treatment, and the thermoelectric properties of the thermoelement branches. To prepare the branches, the following initial materials were used: Bi of TsMTU-3098-52 grade, Te of first grade TsMTU-42-11 (sublimated), Sb of SU-0 grade, and Se of "reagent" grade. Admixtures of 3% Te and 1% Pb were used to prepare the positive alloy. Admixtures of 0.5% Bi and 0.06% Cu were used to make the negative alloy. The powder grain dimensions were 3-0.1 mm. The heat treatment duration was varied from 3 to 48 hr. The pressure applied in briquetting varied from 1 to 10 tons/cm². To determine the optimum value of each of the technological parameters of the preparation, thermoelement branches were prepared keeping all but one parameter constant.

The effect of pressure on the electrical properties and the density of the positive branch is shown in Fig. 1. This figure indicates that the highest value of the power coefficient $\alpha^2\sigma$ is reached at a pressure of 8-9 tons/cm² at the highest value of α. The positive branches were subject to 8-hr annealing at 385°C and

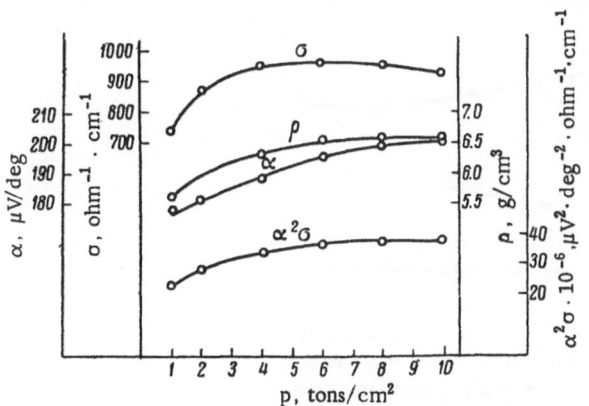

Fig. 1. Effect of pressure on the electrical properties and density of the positive branch alloy $Bi_2Te_3 - Sb_2Te_3$.

79

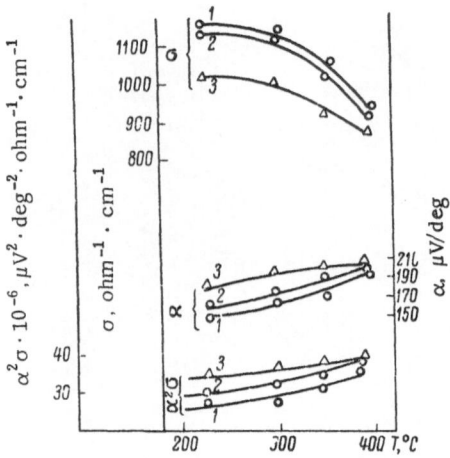

Fig. 2. Effect of the temperature and duration of heat treatment on the electrical properties of the positive branch alloy $Bi_2Te_3 - Sb_2Te_3$. 1) 3-hr heat treatment; 2) 8-hr heat treatment; 3) 48-hr heat treatment.

were pressed from a powder of grain diameter smaller than 0.25 mm.

It is interesting to note that the highest power coefficient occurs at the highest density.

In the present case the power also gives a measure of the efficiency of the branches since the reduction of the density affects the power coefficient of the material more strongly than its thermal conductivity.

The effect of the temperature and duration of the heat treatment on the properties of $Bi_2Te_3 - Sb_2Te_3$ is shown in Fig. 2. Here we see that the power coefficient rises with increase in temperature up to 400°C. Above this temperature the samples are deformed.

Increase of the duration of annealing at temperatures up to 350°C also increases the power coefficient when α rises. However, at an annealing temperature of 390°C heat treatment of more than 8 hr duration has little further effect.

The increase of α of the positive samples with increase of the pressure, temperature, and duration of annealing is due to the presence of Te in the form of a second phase [1]. On increase of the pressure, temperature, and duration of heat treatment, Te enters the crystal lattice and produces negative carriers. This increases the thermoelectric power and reduces the electrical conductivity.

The influence of the initial powder grain size on the thermoelectric properties of $Bi_2Te_3 - Sb_2Te_3$ is illustrated in Fig. 3. This figure shows that with reduction of the grain dimensions the thermoelectric power increases, the electrical conductivity decreases, the lattice thermal conductivity κ_l decreases, and the value of z increases. The increase of z occurs on reduction of the grain dimensions to less than 0.25 mm. This increase is obviously due to the in-

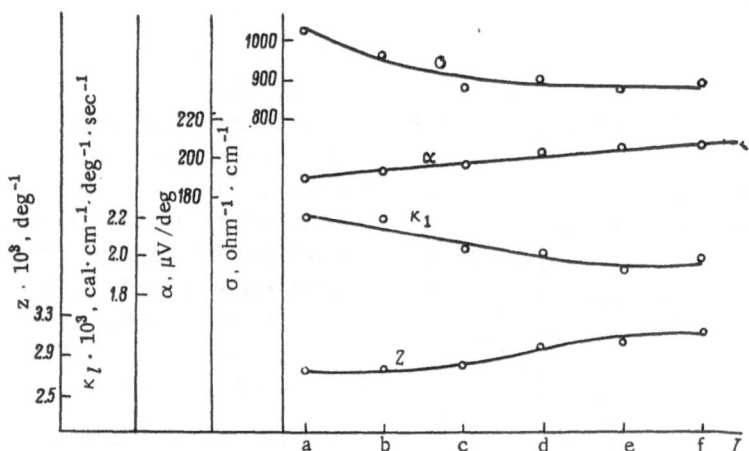

Fig. 3. Effect of the grain dimensions on the thermoelectric properties of the positive branch alloy $Bi_2Te_3 - Sb_2Te_3$. The letters along the abscissa axis represent the following ranges of the grain dimensions d (in mm): a) $2 < d < 3$; b) $1 < d < 2$; c) $0.5 < d < 1$; d) $0.25 < d < 0.5$; e) $0.2 < d < 0.25$; f) $0.1 < d < 0.2$.

Effect of the Grain Dimensions on the Thermoelectric Properties of the Nega-tive Branch Alloy $Bi_2Te_3 - Bi_2Se_3$

Sample No.	σ, ohm⁻¹·cm⁻¹	α, μV/deg	$\alpha^2\sigma \cdot 10^{-6}$, μV²·deg⁻²·ohm⁻¹·cm⁻¹	κ_l, cal·cm⁻¹·deg⁻¹·sec⁻¹	$z \cdot 10^3$, deg⁻¹
319	960	—164	25.8	1.9	2.2
320	930	—161	24.1	1.9	2.0
321	950	—170	27.5	1.8	2.4
322	940	—164	25.3	1.8	2.2
323	920	—170	26.6	1.7	2.3
96	1200	—145	25.2	2.2	1.78
96	1220	—142	24.6	2.2	1.73
96	1220	—146	26.0	2.2т	1.82

Note: Grain size was 2 mm for sample 96 and 0.25 mm for other samples.

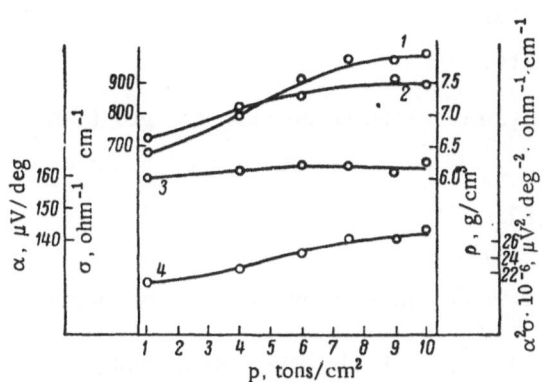

Fig. 4. Effect of the pressure on the electrical prop-erties and density of the negative branch alloy $Bi_2Te_3 - Bi_2Se_3$. 1) σ; 2) ρ; 3) α; 4) $\alpha^2\sigma$.

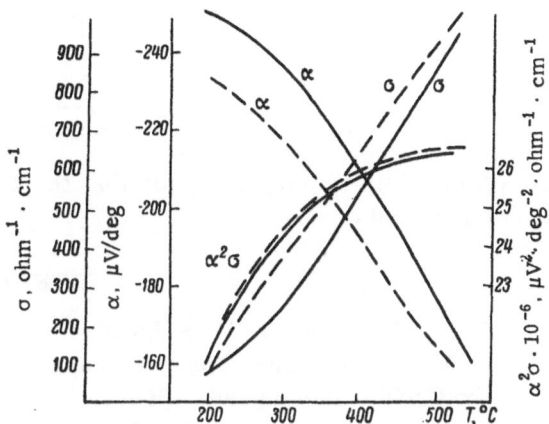

Fig. 5. Effect of the temperature and duration of heat treatment on the electrical properties of the nega-tive branch alloy $Bi_2Te_3 - Bi_2Se_3$. The dashed curves represent an 8-hr heat treatment; the continuous curves represent a 24-hr heat treatment.

crease of the intensity of the interaction between the grains on reduction of their dimensions, i.e., due to the increase of the uniformity of the $Bi_2Te_3 - Sb_2Te_3$. (The measurements of the thermal conductivity were carried out by A. F. Ioffe.)

The effect of the pressure on the electrical properties of the negative branches is shown in Fig. 4. (The samples were subjected to 8-hr annealing at 530°C and were pressed from a powder of grain diameter smaller than 0.25 mm.) As for the positive branches, it was found that a pressure of 8-9 tons/cm² is the optimum value.

The dependence of the electrical properties of the negative branch on the temperature and duration of annealing is illustrated in Fig. 5. With increase of the annealing temperature the electrical conductivity increases but the thermoelectric power decreases. Moreover, the power coefficient increases. Consequently the figure of merit of the alloy is practically con-stant between 400 and 530°C. Above 530°C the negative alloy samples become deformed. In-crease of the duration of heat treatment to more than 8 hr has practically no effect on the elec-trical properties of the branches.

One of the main problems is the stability of the $Bi_2Te_3 - Bi_2Se_3$ alloy [3]. The highest stability of the negative branch was obtained at the highest temperature of heat treatment (510-530°C).

The dependence of the thermoelectric properties of the $Bi_2Te_3 - Bi_2Se_3$ alloy on the grain dimensions is given in the table. The table shows that, as in the case of the positive alloy, the value of z increases with decrease of the grain dimensions.

Using positive and negative materials obtained under optimum conditions we prepared thermoelements for which the value of ΔT_{max} was measured for T_h = 296-298°K. The cold junction of the thermoelement was covered with cotton wool. We thus obtained ΔT_{max} = 55-56° and z = $1.92 \cdot 10^{-3}$ deg^{-1}. The value of ΔT_{max} obtained for samples which were nearly single-crystal in structure was 58° for T_h = 295°K; in this case z was $2.06 \cdot 10^{-3}$ deg^{-1} [4].

In constructing thermopiles it is necessary to known the mechanical strength of the thermo-element branches. We give here the ultimate compressive strength (in kg/mm^2) for samples of the positive alloy $Bi_2Te_3 - Sb_2Te_3$ prepared by hot and cold pressing.

Hot-pressed samples	Cold-pressed samples
4.69	5.95
3.23	4.52
4.15	4.84
4.96	4.33
5.23	5.28
Average 4.46	4.98

From these data we may conclude that the cold-pressed samples are somewhat stronger than the hot-pressed ones.

LITERATURE CITED

1. S. S. Sinani and G. V. Kokosh, Fiz. Tverd. Tela 2:1118, 1960.
2. S. V. Airapetyants, Dissertation for Candidate's Degree, Inst. Poluprovodnikov Akad. Nauk SSSR, Leningrad, 1960.
3. G. N. Gordyakova, G. V. Kokosh, and S. S. Sinani, Zhur. Tekh. Fiz. 28:3, 1958.
4. A. D. Goletskaya, V. A. Kutasov, and E. A. Popova, Fiz. Tverd. Tela 3:3003, 1961.

PREVENTION OF AGING OF NEGATIVE THERMOELEMENT BRANCHES

A. N. Voronin, R. Z. Grinberg, and A. P. Savel'eva

The alloy $Bi_2Te_3 - Bi_2Se_3$ with Bi and Cu admixtures is used for the negative branches of thermoelectric cooling devices [1].

Timed tests under operating conditions have shown that samples prepared from this alloy with the admixture of copper exhibit marked variations of the electric properties with time (the electrical conductivity decreases and the thermoelectric power increases) which increase the powder consumed by the cooling couple and lower its figure of merit z. Measurements were carried out on polycrystalline samples ($20 \times 5 \times 5$ mm) pressed in a cold mold and then heat-treated [1, 2]. The variation of the electrical conductivity σ and the thermoelectric power α with time is given in Fig. 1. The measurements carried out show that in samples prepared by the "cold" pressing method the electrical conductivity falls by 30% in two months and this is accompanied by a $\sim 18\%$ increase of the thermoelectric power.

The most likely cause of the change in the electrical properties of negative branches is their oxidation. To prove this the following experiment was carried out. Cold-pressed samples were subjected, after briquetting, to heat treatment in evacuated ampoules (10^{-1}-10^{-2} mm Hg). After heat treatment, samples prepared from the same alloy were placed in two separate ampoules. Both these ampoules were subjected to the same heat treatment. Measurements on samples from one of the ampoules were carried out immediately on completion of the heat treatment, while the samples in the other ampoule were left in vacuum. After one month, measurements were made on the samples from both ampoules. The results of the measurements, listed in Table I, show that the properties of the samples stored in the sealed ampoule were practically unaffected. Consequently the properties of the other sample could have been altered only by the ambient atmosphere. These results lead to the natural conclusion that oxy-

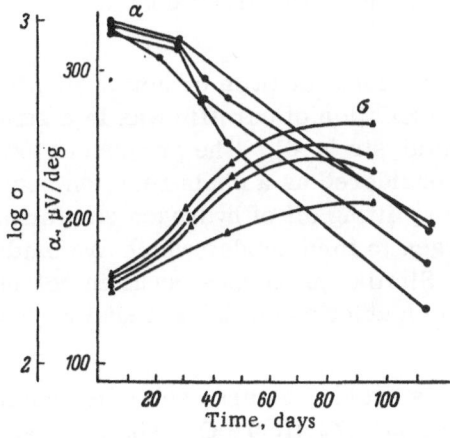

Fig. 1. Variation of the electrical conductivity σ and thermoelectric power α with time for four different samples.

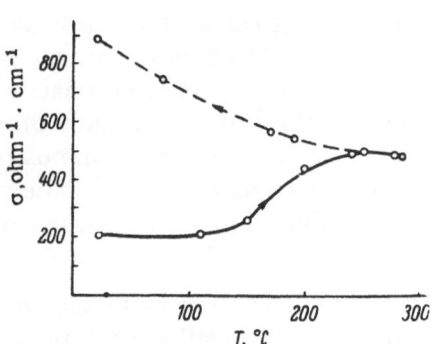

Fig. 2. Effect of chemical reduction in paraffin wax on the electrical conductivity σ of a negative alloy sample.

Table I. Change in the Electrical Properties of Samples

σ, ohm^{-1}·cm^{-1}	α, μV/deg	σ, ohm^{-1}·cm^{-1}	α, μV/deg
		Stored in vacuum	
—	—	1260	—157
—	—	970	—176
—	—	950	—188
—	—	1000	—175
—	—	990	—174
—	—	905	—181
—	—	860	—186
—	—	900	—176
		Not stored in vacuum	
1025	—166	720	—204
970	—160	630	—207
990	—170	610	—212
1010	—168	700	—208

Note. The duration of aging of all the samples was 26 days.

gen plays a direct part in the mechanism of the change of the electrical conductivity and the thermoelectric power.

Confirmation of the important role of oxygen in the change of the electrical properties of semiconductors is also provided by experiments on the protection of samples from the action of oxygen by coating. We used the following materials for this purpose: paraffin wax, a mixture of paraffin wax and colophony, bitumen, and a mixture of bitumen and paraffin wax. (The choice of a particular protective coating composition was determined by the temperature of the hot junction of the thermopile.) All these materials possess, apart from their protective properties, reducing properties as well, since they are high-molecular organic compounds which decompose on heating into low-molecular compounds capable of oxidation. Reduction may occur only in the samples which change their properties near room temperature. The samples heated in the presence of atmospheric oxygen are either poorly reduced or not reduced at all.

The general picture of the action of ambient oxygen on the ternary alloy may be as follows. A sample adsorbs oxygen on its surface. This oxygen reacts with the copper which is present in the surface layers of the sample, forming a solid solution with $Bi_2Te_3 - Bi_2Se_3$; copper, by virtue of the mass law, begins to diffuse from the interior to the surface in order to equalize the volume concentration. At the surface this copper also reacts with oxygen and therefore the concentration of free copper in the sample decreases. The compound of copper and oxygen decomposes easily on heating in a medium capable of removing or reacting with oxygen. This part of the process may be regarded as reversible. As the temperature is increased, the amount of adsorbed oxygen decreases and the copper oxidizes to CuO. As a result of this oxidation the electrical conductivity is reduced to zero. The compound CuO is stable. This part of the process may be regarded as irreversible [3].

The reducing action of hydrocarbons may be considered using paraffin wax as an example. Paraffin wax is a mixture of saturated hydrocarbons with carbon content varying from C_{16} to C_{30}.

At room temperature, oxygen and strong oxidants react but little or not at all with paraffin hydrocarbons. It has been established that strong oxidation of paraffin wax in a stream of air begins from 100-115°C in the presence of metal catalysts [4-6]. The process of oxidation of paraffin wax in the presence of a catalyst may be considered as a chain reaction accompanied by the breaking of carbon atom chains, with the splitting off of hydrogen and formation of radicals. The radicals are easily oxidized by oxygen to form acids, which are oxidized further to form CO_2, H_2O, and low-molecular acids. Similar processes occur in colophony and bitumen. Colophony consists of resin acids of which abietic acid is best known, having the composition $C_{20}H_{30}O_2$.

Bitumen contains high-molecular organic alcohols, acids, esters, and other compounds, corresponding to the overall percentage composition 88% C, 7% H, 5% O. The process of protection of a negative-alloy sample is illustrated by reduction of a sample in paraffin wax at t = 150-250°C (Fig. 2).

Table II. Results of an Investigation of Aging of Samples Covered with Protective Layers

Type of layer	Duration of storage, days	Aging,%	Remarks
Bitumen and paraffin wax	188	0	Encapsulated in mixture
	135	0	
	45	0	
Bitumen	22	0	Encapsulated in bitumen
	75	0	
Colophony and paraffin wax	112	0	Encapsulated in mixture
	2.5 years	0	Coated with thin layer
Paraffin wax	40	0	Coated with thin layer

Table III. Results of an Investigation of Aging of Samples Covered with Disturbed Protective Layers

Type of layer	Duration of storage, days	Aging, %	Remarks
Bitumen	260	6.38	Coated with thin layer. Edges exposed.
Bitumen and Paraffin	260	0	Coated with thin layer. Edges covered.
	260	2.4	Coated with thin layer. Edges exposed.

After treatment, the samples coated with a layer of paraffin wax exhibited no change of their electrical properties with time (Table II). Similar behavior has also been observed on treatment of samples in bitumen, in a mixture of bitumen and paraffin wax, and in a mixture of colophony and paraffin wax. The samples coated with these materials or with a continuous thin film do not age.

When the protective film is distubed, the properties of the samples alter by about 10 % per year (Table III). To avoid this, the sample edges should either be covered or the film thickness should be increased, though the latter procedure is undesirable. Table III lists the results of measurements on samples which were subjected to the same treatment in a mixture of bitumen and paraffin wax, but in some cases the edges were sharp and exposed and in others they were covered with a continuous protecting layer $\sim 300 \ \mu$ thick.

The samples with covered edges exhibited no change in their electrical properties. It should be noted that heat treatment was necessary before coating the samples with any of the three materials. The temperature of the heat treatment was 80-200°C and its duration was 15-20 min.

The best results were obtained at the following temperatures: 200°C in bitumen, 180°C in a mixture of bitumen, and paraffin wax, and 150°C in a mixture of colophony and paraffin wax or in pure paraffin wax.

The need for heat treatment before coating with reducing materials is due to the fact that the cold-pressed samples have a porosity amounting to 3-4 % by volume. The pores contain air which is expelled by the coating material in its liquid state. Therefore simple deposition of the protective coatings by solidifying them on the surface is insufficient.

LITERATURE CITED

1. A. N. Voronin and R. Z. Grinberg, present collection, p. 79.
2. S. S. Sinani, G. V. Kokosh, and G. N. Gordyakova, Zhur. Tekh. Fiz. 28:3, 1958.
3. B. T. Kolomiets and V. I. Larichev, Zhur. Tekh. Fiz. 28:1358, 1958.
4. B. N. Dolgov, Catalysis in Organic Chemistry, Goskhimizdat, Moscow-Leningrad, 1948.
5. A. E. Chichibabin, Basic Introduction to Organic Chemistry, Goskhimizdat, Moscow-Leningrad, 1953.
6. General Chemical Technology, ed. S. I. Vol'fkovich, Goskhimizdat, Moscow-Leningrad, Vol. 1, 1952; Vol. 2, 1959.

DOMESTIC THERMOELECTRIC REFRIGERATOR OF 20 LITER CAPACITY

A. N. Voronin, S. G. Platonova, E. G. Pokornyi, and É. M. Sher

In developing a domestic thermoelectric refrigerator, the main difficulty is the selection of the optimum system for heat removal from the "hot" junctions of the semiconductor thermoelectric battery. On the one hand it is always convenient to have the dimensions of the thermoelectric battery at a minimum, and on the other it is desirable to have the surface of the "hot" junctions as large as possible to ensure the best heat exchange between them and the surrounding medium. In the best samples of thermoelectric refrigerators, in which the heat is removed from the "hot" junctions by means of a large number of fins, the temperature drop between the "hot" junctions and the ambient medium is not less than 12°. This makes it necessary to increase the working temperature drop across the thermopile and therefore to reduce the coefficient of performance. Consequently the coefficient of performance of a domestic thermoelectric refrigerator using natural convection of air for the removal of heat from the "hot" junctions does not exceed 25-27%. Moreover, the weight of the fins is then quite considerable.

A marked reduction of the temperature drop on the "hot" side of the thermopile and the weight of the fins may be obtained by a splash-evaporation method of heat removal from the "hot" junctions. A device using this method is shown in the figure.

A tank 1 contains water which is gravity fed to the working chamber of a pump 5 through the jacket of the thermopile 6 which contains the fins of the "hot" junctions. The pump pushes the water through a splash nozzle 3, forcing it to rise in the form of a fountain. The pump is operated by a small (4 W) electric motor without a commutator. Some of the water rising as a fountain evaporates and this ensures heat transfer from the "hot" junctions to the ambient medium. Under steady-state conditions the temperature of the "hot" junctions is kept at the level of the ambient temperature (when the temperature of the ambient atmosphere is 20°C, the temperature of the "hot" junctions amounts to 20-21°C). Therefore the coefficient of performance of the thermopile of the refrigerator is about 50%.

For a refrigerator with a useful volume of 20 liters the amount of water consumed by evaporation is about 0.5 liters/day. The tank 1 has a capacity of 3.5 liters and needs refilling not more than once or twice a week.

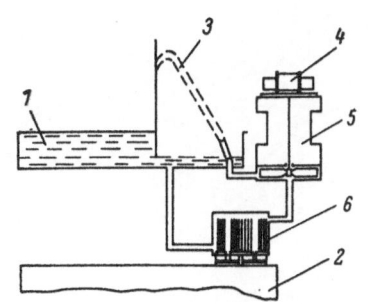

Schematic representation of the refrigerator.

The thermopile of the 20-liter refrigerator has eight thermoelements connected in series and having branches (arms) with dimensions $S = 6.5 \times 7$ mm, $l = 5$ mm; the weight of the semiconducting material is 30 g. The working current of the thermopile is 20 A. The power consumed by the battery is 10 W, the weight of the battery (which includes the semiconducting material and the fins of the "hot" junctions) is 250 g, and its dimensions are $40 \times 40 \times 40$ mm. The battery is soldered to the top of the working chamber of the refrigerator 2 via two electrically insulated bridging plates.

The battery is supplied from a full-wave rectifier, each arm of which includes a germanium diode of VG-10-15 type.

The choke of the rectifier filter has about 500 ampere-turns. If the water supply becomes exhausted, the refrigerator is disconnected from the mains by means of a thermal cut-out (a bimetal plate) soldered to the jacket of the battery.

Since the working voltage of the battery is the same as the voltage drop across the germanium diodes (0.5 V), the rectifier efficiency is low. Therefore the power taken from the a-c mains by the refrigerator amounts to 30 W. The efficiency of the rectifier may be improved by increasing the voltage drop across the thermopile (i.e., by reducing the working current and increasing the number of thermoelements).

The rectifier is connected to 127-V or 220-V a-c mains and is made in the form of a unit which is built into the refrigerator casing. Foamed polystyrene 10 cm thick serves as the thermal insulation of the refrigerator.

The temperature at the center of the working chamber of the refrigerator is lowered by 20° relative to the ambient temperature. The time for the refrigerator to reach its steady state is about 3 hr.

The method described here for removing heat from the "hot" junctions of the battery is suitable for refrigerators with a useful volume up to 30 liters. For larger volumes of the working chamber it would be necessary to evaporate large amounts of water, which makes this method impracticable.

METHOD OF PREPARING STABLE n-TYPE MATERIAL
FOR THERMOELEMENTS WORKING UNDER COOLING CONDITIONS

D. I. Lainer, L. M. Ostrovskaya, M. G. Épshtein,
V. I. Kochkarev, and V. P. Khavricheva

The ternary alloys bismuth – tellurium – selenium (for the negative arms of thermo-elements) and bismuth – tellurium – antimony (for the positive arms) are currently widely used in the industrial production of cooling devices. The composition of the alloys and the technique of their manufacture have been developed fairly recently at the Institute for Semiconductors of the USSR Academy of Sciences.

The Bi – Te – Se alloy is a solid solution of the two intermetallic compounds Bi_2Te_3 and Bi_2Se_3 in proportions of 80 and 20 mol.% [1]. The properties of this solid solution are satisfactory in practice only after the introduction of small amounts of alloying admixtures which raise considerably the electrical conductivity of the alloy. Extensive investigations at the Institute of Semiconductors have established that Cu, Ag, $CdBr_2$, and Hg_2Cl_2 are effective admixtures, but the alloys with Cu or Ag are unsuitable because of their low stability. The recommended admixture of Hg_2Cl_2 ensures good and stable properties but leads to difficulties under factory conditions, requiring in particular that melting be carried out in evacuated ampoules made of refractory Pyrex glass or quartz.

The purpose of the present work was to investigate the effect on the n-type ternary alloy Bi – Te – Se, of the composition given above, of several alloying elements, and to find ways of simplifying the technological procedure for the manufacture of the alloy in large quantities.

Alloys containing small amounts of the alloying elements were prepared by melting together the components in evacuated quartz ampoules at 700-720°C. The properties were measured using rectangular pressed samples (semi-elements) by the usual compensation method at room temperature. The influence of several rare elements, including Ga, Nd, Re, and In, as well as of the halides $ZnCl_2$, Hg_2Cl_2, NaCl, and KCl, was investigated. Within the concentration limits of 0.05-0.2 wt.%, alloying with the rare elements gave no promising results.

Of the halides, $ZnCl_2$ constituted a very effective admixture which increased the electrical conductivity of the alloy by a large factor. By introducing small amounts (0.1-0.15 wt.%) of $ZnCl_2$ (Fig. 1) it was possible to obtain an alloy for which the ratio of the electrical conductivity and thermoelectric power was suitable for practical applications: $\sigma = 850$ ohm$^{-1} \cdot$ cm^{-1}, $\alpha = 180-190$ μV/deg, $\alpha^2 \sigma = 30 \cdot 10^6$ μV$^2 \cdot$ deg$^{-2} \cdot$ ohm$^{-1} \cdot$ cm^{-1}. Further increase of the $ZnCl_2$ content in the alloy produced a continuous increase of the electrical conductivity and decrease of the thermoelectric power right up to 1 wt.% of the admixture; at this concentration of $ZnCl_2$ there are discontinuities in the σ, α, and $\alpha^2 \sigma$ curves, which are related to the limit of solubility of $ZnCl_2$ in the alloy.

It is interesting to compare these results with those on the effect of calomel on the properties of the n-type ternary alloy, since calomel has up to now been regarded as the most effective admixture. The effect of calomel was investigated up to a concentration of 0.4 wt.% (Fig. 1). With this amount of calomel the alloy has $\sigma = 2300$ ohm$^{-1} \cdot$ cm^{-1} and $\alpha = 29-30$ μV/deg.

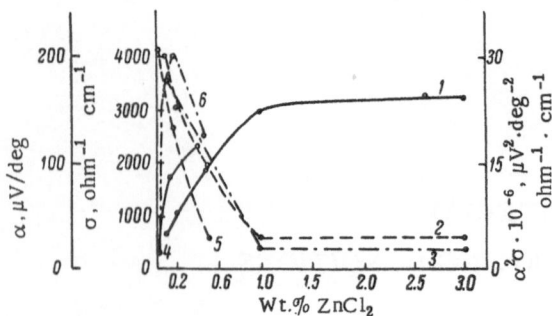

Fig. 1. Dependence of the properties of the ternary n-type alloy on the $ZnCl_2$ or Hg_2Cl_2 content. 1-3) σ, α, and $\alpha^2\sigma$ for $ZnCl_2$; 4-6) σ, α, and $\alpha^2\sigma$ for Hg_2Cl_2.

The value of $\alpha^2\sigma$ reaches a maximum of 30-32 $\cdot 10^6$ $\mu V/deg^{-2} \cdot ohm^{-1} \cdot cm^{-1}$ at 0.2 wt.% of calomel and decreases sharply above this concentration.

The effects of NaCl and KCl on the properties of the n-type alloy were investigated up to 0.4 wt.% admixture. According to our data these halides are insoluble in the n-type alloy. This is supported by the results of an indicator reaction for establishing the presence of chlorine ions and the constancy of the electrical properties of alloys containing different amounts of the admixture. The insolubility of NaCl and KCl in the alloy led us to develop a method of introducing $ZnCl_2$ from a protective flux containing sodium and potassium chlorides.

A flux consisting of equal parts by weight of KCl, NaCl, and $ZnCl_2$ has a melting point of ~400-450°C and until the moment of melting of the main alloy forms a continuous liquid layer which prevents volatilization almost completely. The flux does not mix with the metal and it is possible to obtain a monolithic ingot free from flux occlusions using a demountable heated mold. The necessary degree of alloying is obtained by using 400 g of the flux per 2 kg of the melt; the resulting alloy exhibits the following properties: $\sigma = 800-850$ $ohm^{-1} \cdot cm^{-1}$, $\alpha = 180-190$ $\mu V/deg$, $\alpha^2\sigma = 30 \cdot 10^6 \mu V^2 \cdot deg^{-2} \cdot ohm^{-1} \cdot cm^{-1}$. There is no difficulty in preparing large quantities of the melt because the necessary amount of the flux is easily controlled and an alloy with good properties can be obtained.

This method of alloy manufacture does not require the use of vacuum. Moreover it considerably simplifies the process and reduces the cost makes it possible to prepare large amounts of the alloy in a single melt, and does not require any further alloying. This method is being used successfully at the "Termoelektrogenerator" (Thermoelectric Generator) works.

Special attention was paid to the problem of the stability of the alloy with the $ZnCl_2$ admixture. Tests carried out over 200-220 days on alloys containing $ZnCl_2$ or Hg_2Cl_2, and also on alloys melted under flux and stored in air at 200°C, showed that all these alloys are highly stable (Fig. 2). The electrical conductivity decreased during this period by not more than 1-3%. The stability of the alloys with $ZnCl_2$ was not poorer than the stability of the alloys with calomel. Moreover, the temperature at which the alloys were stored and the free access of

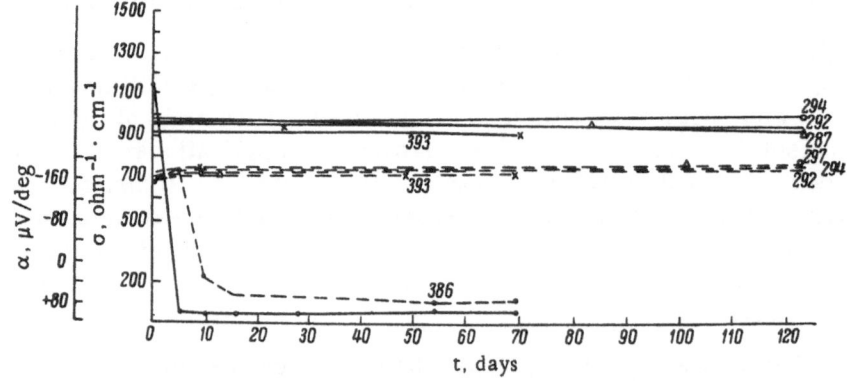

Fig. 2. Variation of the properties of ternary alloys containing admixtures with the duration of annealing at 200°C. The dashed lines indicate the variation of α; the continuous lines give the variation of σ; No. 393 is an alloy with 0.2 wt.% Hg_2Cl_2; Nos. 292, 294 and 297 are alloys with $ZnCl_2$; No. 386 is an alloy with 0.2 wt.% Cu.

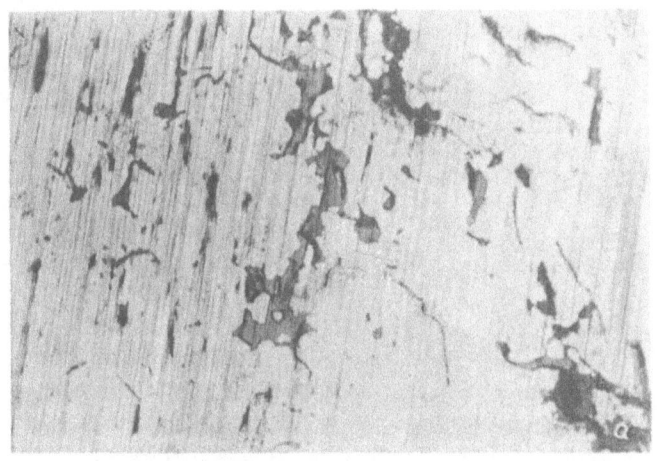

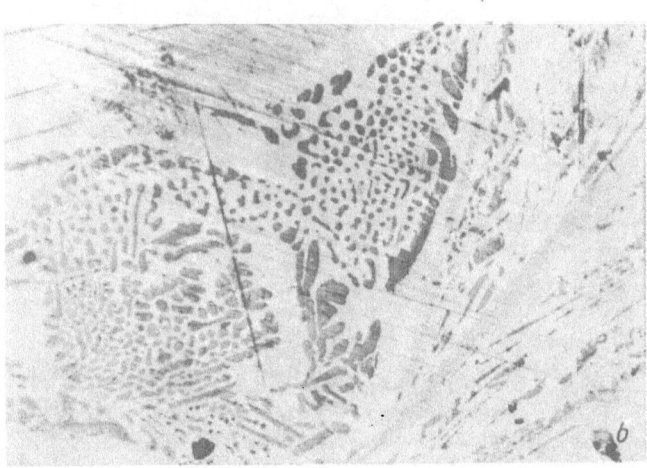

Fig. 3. Microstructure of alloys containing different amounts of copper: a) 0.2 wt.% Cu; b) 5 wt.% Cu. The sections were not etched. × 200.

air did not affect greatly the properties of the alloys with $ZnCl_2$; the nature of the change in the properties of the thermoelements Nos. 292, 294, and 297 is practically identical. The results indicate that the admixture in these alloys does not undergo any changes with time. Figure 2 shows the behavior of an alloy with 0.2% copper prepared without an excess of bismuth in the charge. The electrical conductivity of this alloy decreased by a factor of 10-15 during the first 10 days of strorage in air at 200°C. During the same period the negative thermoelectric power of this alloy rose slightly; later the absolute value decreased sharply and the power became positive. Further storage under the same conditions produced no additional changes.

Obviously the mechanism of such great changes is complex and is not limited to the oxidation of the copper, which would cancel the whole alloying effect. This is also indicated by the reduction of the electrical conductivity by a factor of 5-6 in those cases when the elements were stored in evacuated ampoules at 200°C, i.e., under conditions which excluded the possibility of oxidation. It is more logical to assume that the reason for the deterioration of the properties of the alloy involves internal changes which are not related to the selective oxidation of the admixture.

From our observations we conclude that during the process of manufacture and storage of the alloy the copper is precipitated from the solid solution in the form of a compound with tellurium, producing a deficiency of the latter. It is known that the properties of the alloy are

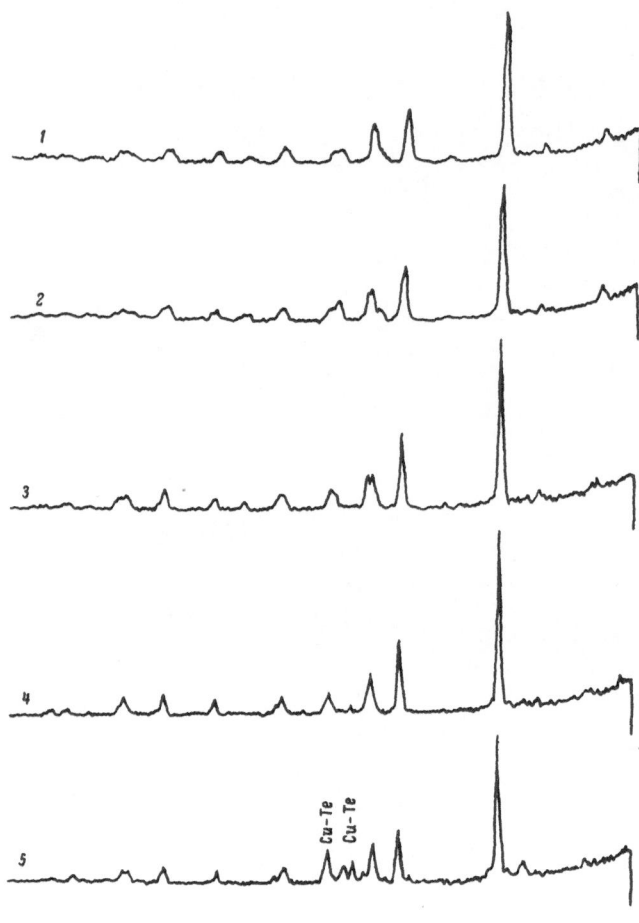

Fig. 4. X-ray diffractograms of alloys containing different amounts of copper;
1) Original Bi − Te − Se alloy; 2) Bi − Te − Se + 0.2% Cu; 3) Bi − Te − Se
+ 5% Cu; 4) Bi − Te − Se + 10% Cu; 5) Bi − Te − Se + 20% Cu.

very sensitive to an excess or deficiency of tellurium, which may even reverse the sign of the thermoelectric power.

The tendency of copper to combine with tellurium was observed in alloys containing from 0.2 to 20 wt.% of copper. The pressed thermoelements have the structure shown in Fig. 3. It can be seen that even in the presence of 0.2 wt.% of copper in the alloy a second phase is precipitated along grain boundaries. As the amount of copper is increased, the amount of the second phase increases correspondingly, but the nature and form of the precipitates remain the same, of eutectic type. Measurements of the microhardness showed that it is always the same phase that is precipitated.

X-ray structure analysis using the apparatus URS-50I established the presence of one of the compounds of the Cu − Te system in alloys with a high copper content. The diffraction patterns (Fig. 4) show lines corresponding to the interference maxima of the compounds Cu_2Te or Cu_4Te_3; these are, however, difficult to distinguish because of the very similar values of their interplanar spacings.

Our hypothesis explains the observation of G. N. Gordyakova and S. S. Sinani that the concentration of copper in the surface layer increases with time.

The present authors know of cases in which a nonuniform distribution of surface-active elements exists in alloys. According to the ideas of Gibbs, developed later by Rebinder, Ar-

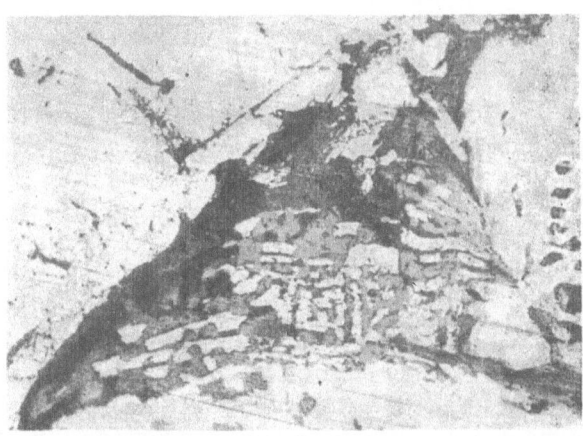

Fig. 5. Microstructure of the alloy with 5 wt.% Cu (traces of oxidation of the second phase are visible). ×200.

kharov, Semenchenko, and others, surface-active elements are concentrated mainly along grain boundaries and on the surface. Copper, however, is the least active of all the elements present in the system under consideration. Therefore we cannot use these ideas to explain our observations. However, if it is assumed that the copper is precipitated from the supersaturated solid solution in the form of a compound with tellurium, the results obtained can be understood: the precipitated copper telluride concentrates along grain boundaries and enriches these boundaries and the surface with copper.

Moreover, we observed on sections of the alloys with copper admixture that copper telluride oxidizes on storage in air for several days. Figure 5 shows a gray background of the second phase with darker oxidized regions. Under a microscope these regions are lilac in color and their microhardness is higher than that of the unoxidized regions. It should also be pointed out that the compound Cu_4Te_3 oxidizes quite strongly even at room temperature, forming a colored oxide layer. It is possible that in the alloys containing copper we are dealing with the compound Cu_4Te_3 and its oxidation.

It seems to us that the mechanism of aging of the alloys containing copper should allow for the ideas put forward above.

Conclusions

1. It was found that zinc chloride is an effective donor admixture which ensures good electrical properties of the n-type bismuth – tellurium – selenium alloy when present in concentrations of 0.1-0.15%.

2. The alloys with zinc chloride prepared under a flux have a high stability which is not less than that of the alloys with calomel.

3. It was established that the volatile admixture of calomel can be successfully replaced with zinc chloride by employing the method of alloying from a protective flux.

4. It is suggested that the low stability of the alloys with copper is due to the precipitation of copper from the solid solution in the form of a compound with tellurium, and thus to the related strong reduction of the carrier density. Oxidation is a secondary process and affects the Cu – Te phase and not copper itself.

LITERATURE CITED

1. G. N. Gordyakova, G. V. Kokosh, and S. S. Sinani, Zhur. Tekh. Fiz. 28:3, 1958.

THERMOELECTRIC COOLING BY MEANS OF HEAT
FROM LOW-GRADE SOURCES

I. N. Pomazanov and P. L. Tikhomirov

The present authors have described earlier [1, 2] the principle of action of a semiconductor electron heat pump (EHP) making it possible to transfer heat from a cooler to a hotter body by using the thermoelectric effects in semiconductors. In this pump the refrigerator is supplied from a thermoelectric generator consisting of two semiconductor plates joined by a metal bridge. To obtain the maximum heat pumping rate the temperature difference between the thermocouple junctions of the generator should be of the order of several hundred degrees.

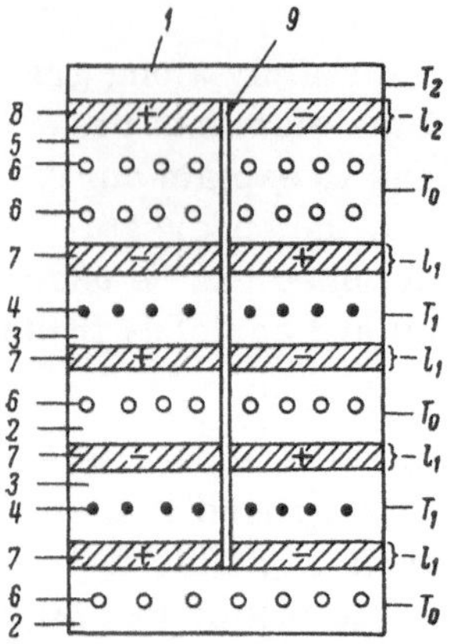

We shall consider the possibility of reducing this temperature difference by using in the generator several thermocouples connected in series so that the emf's of the thermocouples are additive.

Figure 1 shows the basis of the device in the case of a thermoelectric generator consisting of four thermocouples.

Let n be the number of thermocouples making up the thermoelectric generator, and let T_0 be the temperature of plates 2 and 5, T_1 the temperature of plate 3, and T_2 the temperature of plate 1. Then the emf acting in the closed circuit of the device is

$$\varepsilon = a\,(n\Delta T_1 - \Delta T_2), \tag{1}$$

where α is the thermoelectric power relative to a pair of plates; $\Delta T_1 = T_1 - T_0$; $\Delta T_2 = T_0 - T_2$.

Neglecting the resistance of the metal plates and the bridging layers, we obtain the following expression for the current:

$$I = \frac{a\,(n\Delta T_1 - \Delta T_2)\,s}{(\rho + \rho_2)\,(nl_1 + l_2)}, \tag{2}$$

where ρ_1 and ρ_2 are the resistivities of the semiconducting materials of the thermocouples; l_1 is the thickness of the semiconducting plates 7; l_2 is the thickness of the semiconducting plates 8; s is the area of the cross section of the semiconducting plates 7 and 8.

The heat-pumping rate, i.e., the thermal power (Q_2) absorbed by the plate 1 by evolution of the negative Peltier heat, the Joule heat, and heat transfer due to the thermal conductivity of the semiconducting plates, is given by the expression

$$Q_2 = \frac{a^2\,(n\Delta T_1 - \Delta T_2)\,s}{(\rho_1 + \rho_2)\,(nl_1 + l_2)}\left[T_2 - \frac{(n\Delta T_1 - \Delta T_2)\,l_2}{2\,(nl_1 + l_2)}\right] - (\kappa_1 + \kappa_2)\frac{\Delta T_2}{l_2}\,s, \tag{3}$$

Fig. 1. Basic diagram of a thermoelectric cooling device supplied from a thermoelectric generator consisting of four thermocouples. 1) Metal (copper) plate; 2, 5) metal plates kept at the temperature $T_0 < T_1$ by circulating a liquid (water) through the apertures 6; 3) metal plates kept at the temperature T_1 by heat evolution in the apertures 4 of these plates; 7) plates of p-type (+) and n-type (−) semiconductors which make up the generator; 8) plates of p-type (+) and n-type (−) semiconductors which make up the cooling device; 9) insulating gap.

$\frac{l_1}{l_2}$	n	$(\Delta T_1)_{max\,Q_2}$	$z = 2\cdot10^{-3}$, deg^{-1}		$z = 4\cdot10^{-3}$, deg^{-1}		$z = 10\cdot10^{-3}$, deg^{-1}		$z = 25\cdot10^{-3}$, deg^{-1}	
			$(\Delta T_1)_{max\,\xi}$	ε_{max}, %	$(\Delta T_1)_{max\,\xi}$	ε_{max}, %	$(\Delta T_1)_{max\,\xi}$	ε_{max}, %	$(\Delta T_1)_{max\,\xi}$	ε_{max}, %
1	1	546	280	7.7	195	15.6	126	31.7	93	48.4
1	2	405	190	4.0	140	8.4	84	17.9	60	27.2
1	4	334	165	2.7	112	5.5	72	10.8	46	15.3
1	10	291	145	1.2	100	2.6	60	4.6	36	6.5
0.5	1	415	212	6.8	153	13.1	97	26.4	75	43.5
0.5	2	273	130	4.1	100	8.9	70	18.1	47	27.7
0.5	4	207	106	2.7	7.5	5.5	50	8.8	34	13.1
0.5	10	159	92	1.3	58	2.6	36	4.3	28	6.3
2	1	809	413	8.9	269	19.3	167	38.6	104	53.6
2	2	668	320	5.2	205	10.8	123	19.6	80	28.8
2	4	597	290	3.0	193	6.4	112	10.7	64	14.9
2	10	554	270	1.3	175	2.1	100	4.4	58	6.2

Note: $T_0 = 283°K$, $\Delta T_2 = 20°$.

where κ_1 and κ_2 are the thermal conductivities of the semiconducting materials of the thermocouples (the thermal resistances of the metal bridges are neglected).

To keep the plates 3 at the temperature $T_1 > T_0$ a quantity of thermal power Q_1 should be supplied to them. Allowance for the Peltier heat, the Joule heat, and the heat transferred through the semiconducting plates of the generator leads to the expression

$$Q_1 = \frac{a^2(n\Delta T_1 - \Delta T_2)\,sn}{(\rho_1+\rho_2)(nl_1+l_2)}\left[T_1 - \frac{(n\Delta T_1-\Delta T_2)\,l_1}{2(nl_1+l_2)}\right] + (\kappa_1+\kappa_2)\frac{\Delta T_1}{l_1}\,sn. \tag{4}$$

The efficiency of the device is given by the coefficient of performance,

$$\xi = \frac{Q_2}{Q_1} = \frac{z(n\Delta T_1-\Delta T_2)T_2 - \frac{1}{2}z(n\Delta T_1-\Delta T_2)^2\dfrac{l_2}{nl_1+l_2} - \Delta T_2\dfrac{nl_1+l_2}{l_2}}{\left[z(n\Delta T_1-\Delta T_2)T_1 - \frac{1}{2}z(n\Delta T_1-\Delta T_2)^2\dfrac{l_1}{nl_1+l_2} + \Delta T_1\dfrac{nl_1+l_2}{l_1}\right]\cdot n}, \tag{5}$$

where

$$z = \frac{a^2}{(\rho_1+\rho_2)(\varkappa_1+\varkappa_2)}.$$

Considering the heat-pumping rate Q_2 of Eq. (3) and the coefficient of performance of Eq. (5) as functions of the temperature difference ΔT_1 and seeking the maxima of these functions for fixed values of the other parameters of the device, we find the value of $\Delta T_1 = (\Delta T_1)_{max\,Q_2}$ for which the heat-pumping rate is maximum and the value of $\Delta T_1 = (\Delta T_1)_{max\,\xi}$, for which the coefficient of performance is maximum:

$$(\Delta T_1)_{max\,Q_2} = \frac{1}{n}T_0 + T_2\frac{l_1}{l_2}, \tag{6}$$

$$(\Delta T_1)_{max\,\xi} = \frac{b+\sqrt{b^2+4ac}}{2a}, \tag{7}$$

where

$$a = \frac{1}{2}zn^2\left[\frac{T_2(n+1)z}{n\dfrac{l_1}{l_2}+1} + T_2 z + \frac{l_2}{l_1}\right],$$

$$b = zn\Delta T_2\left[\frac{T_0(n+1)z}{n\dfrac{l_1}{l_2}+1} + T_2 z + \frac{l_2}{l_1}n + 2\right],$$

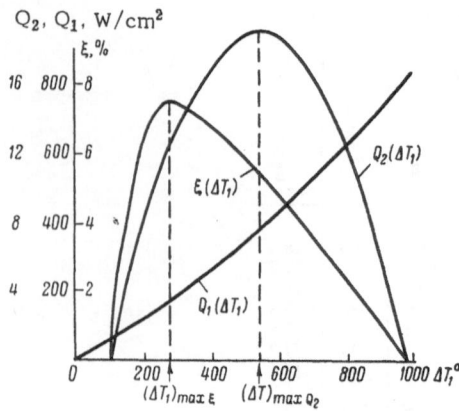

Fig. 2. Dependence of the heat-pumping rate Q_2, thermal power consumed Q_1, and coefficient of performance ξ on the temperature difference between the generator plates.

$$c = z T_2 \Delta T_2 \left(n + \frac{l_1}{l_2}\right) + z \Delta T_2 (n T_0 - \Delta T_2)\left(n \frac{l_1}{l_2} + 1\right) +$$

$$+ z n \Delta T_2^2 \frac{l_1}{l_2} + \frac{1}{2} z \Delta T_2^2 \frac{l_2}{l_1} + \Delta T_2 \left(n^2 \frac{l_1}{l_2} + 2n + \frac{l_2}{l_1}\right) +$$

$$+ \frac{1}{2} \frac{z^2 n \Delta T_2^2}{n l_1 + l_2} (l_1 T_2 - l_2 T_0) - z^2 T_2 \Delta T_2^2 - \frac{1}{2} z^2 \Delta T_2^3 \frac{l^2}{n l_1 + l_2}.$$

From Eqs. (6) and (7) it follows that increase of the number of series-connected thermocouples in a thermoelectric generator reduces the values of $(\Delta T_1)_{\max Q_2}$ and $(\Delta T_1)_{\max \xi}$ which, moreover, depends on the ratio of the semiconducting plate thicknesses and on the temperatures T_0 and T_2. The value of $(\Delta T_1)_{\max Q_2}$ is independent of the physical properties of the semiconducting materials, but the value of $(\Delta T_1)_{\max \xi}$ depends on the figure of merit z of the semiconducting materials.

The values of $(\Delta T_1)_{\max Q_2}$ and $(\Delta T_1)_{\max \xi}$ calculated by means of Eqs. (6) and (7) for some values of n, l_1/l_2, T_2, T_0, and z are listed in the table. The table also gives the maximum value of the coefficient of performance $\xi_{\max}$ corresponding to the values $(\Delta T_1)_{\max Q_2}$.

From the table it follows that both $(\Delta T_1)_{\max Q_2}$ and $(\Delta T_1)_{\max \xi}$ may vary within wide limits depending on the values of the other parameters. The maximum coefficient of performance is displaced along the temperature scale toward lower values of ΔT_1 than the position of the maximum heat-pumping rate on the temperature scale. To illustrate this shift of the maxima, Fig. 2 gives curves calculated using Eqs. (3)-(5) for the case z = $2 \cdot 10^{-3}$ deg^{-1}, n = 1, T_0 = 283°K, ΔT_2 = 20°, $\alpha = 4 \cdot 10^{-4}$ V/deg, $\rho_1 = \rho_2 = 10^{-3}$ ohm $\cdot$ cm, $\kappa_1 = \kappa_2 = 5 \cdot 10^{-3}$ cal $\cdot$ cm$^{-1} \cdot$ sec$^{-1} \cdot$ deg^{-1}, $l_1 = l_2 = 0.1$ cm. From the curves it follows that the conditions for the maximum heat-pumping rate are reached at $(\Delta T_1)_{\max Q_2}$ = 546° while the maximum coefficient of performance is reached at $(\Delta T_1)_{\max \xi}$ = 280°.

The presence of a broad maximum on the $Q_2 = f(\Delta T_1)$ curve leads to the fact that on transition from $(\Delta T_1)_{\max \xi}$ to $(\Delta T_1)_{\max Q_2}$ the quantity Q_2 varies slightly but $Q_{2\,\max}$ increases strongly. Consequently, the heat-pumping rate under the conditions of maximum efficiency differs only slightly from $Q_{2\,\max}$, but the coefficient of performance at the maximum heat-pumping rate is considerably smaller than $\xi_{\max}$. Thus the condition for maximum heat-pumping rate is not optimal from the energy point of view. On the other hand, the condition for maximum efficiency (maximum coefficient of performance) is not only rational from the point of view of the efficiency itself, but also from the point of view of the increase of the service life of the device, since under these conditions the semiconducting plates of the generator are at a lower temperature than under the conditions for the maximum heat-pumping rate.

Figure 3 gives curves showing the displacement of $(\Delta T_1)_{\max \xi}$ and $(\Delta T_1)_{\max Q_2}$ when the value of ΔT_2 is varied. These curves were calculated using Eqs. (6) and (7) for the same values of $\alpha, \rho, \kappa, l_1, l_2$, n, T_0 as above. From the curves of Fig. 3 it is clear that the difference $(\Delta T_1)_{\max Q_2} - (\Delta T_1)_{\max \xi}$ decreases with increase of T_2, reaching zero at $\Delta T_2 = T_{2\,\max}$.

The table shows that the value of $(\Delta T_1)_{\max \xi}$ decreases on increase of n and on decrease of the ratio l_1/l_2, and may amount to only several tens of degrees. For example, $(\Delta T_1)_{\max \xi}$ = 36° when z = $4 \cdot 10^{-4}$ deg^{-1}, ΔT_2 = 20°, T_0 = 283°K, l_1/l_2 = 0.5, and n = 10. Thus it becomes possible to obtain a cooling effect using low-grade heat sources.

The value of the coefficient of performance increases with increase of the figure of merit of the semiconducting materials z. At present, the materials used in industry have z = $2 \cdot 10^{-3}$ deg^{-1}, which corresponds to the coefficient of performance $\xi_{\max}$ = 7-8% (for ΔT_2 = 20°,

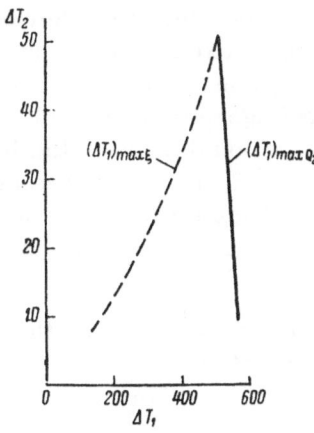

Fig. 3. Displacement of the difference between the temperatures of $(\Delta T_1)_{\max \xi}$ and $(\Delta T_1)_{\max Q_2}$ on variation of ΔT_2.

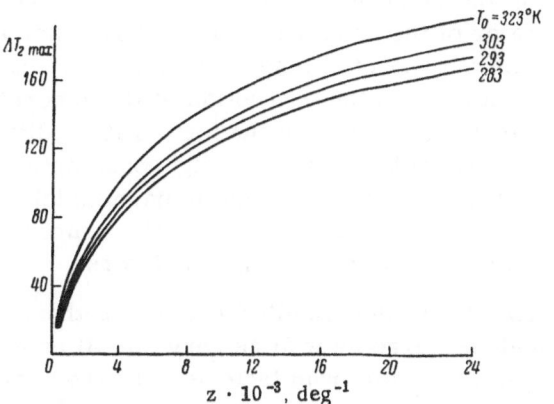

Fig. 4. Maximum possible values of ΔT_2 as a function of z for a fixed value of T_0.

$T_0 = 283°K$, $n = 1$, $l_1/l_2 = 1$). However, there have been reports [3, 4] of the real possibilities of using in the immediate future materials with figures of merit of the order of $z = 3\text{-}4 \cdot 10^{-3}$ deg^{-1}, which should make it possible to increase the value of $\xi_{\max}$ to about 15%.

Substituting the value $(\Delta T_1)_{\max Q_2}$, given by Eq. (6) into Eq. (3), we find an expression for the maximum heat-pumping rate $Q_{2\,\max}$:

$$Q_{2\,\max} = \frac{a^2 T_2^2 s}{2(\rho_1 + \rho_2) l_2} - (\kappa_1 + \kappa_2) \frac{\Delta T_2}{l_2} s. \tag{8}$$

Consequently, the maximum heat-pumping rate $Q_{2\,\max}$ at a given value of ΔT_2 is determined by the physical properties of the semiconducting plates of the generator, the geometrical dimensions of these plates, and the temperature T_0, but it is independent of the number of thermocouples in the generator.

The maximum possible temperature drop in the cooling device is found from Eq. (8) under the condition of zero thermal load, i.e., $Q_{2\,\max} = 0$:

$$\Delta T_{2\,\max} = \frac{1}{2} z T_2^2 \tag{9}$$

or

$$T_{2\,\min} = T_0 - \frac{1}{z}(\sqrt{2z T_0 + 1} - 1). \tag{10}$$

The relationships (9) and (10) are valid for any thermoelectric cooling device irrespective of the method of power supply [5]. From Eq. (10) it follows that $\Delta T_{2\,\max} = T_0 - T_{2\,\min}$ is governed by the figure of merit z of the semiconducting materials used in the cooling device and by the temperature T_0. This relationship is illustrated by curves in Fig. 4 calculated using Eq. (10) for several particular values of T_0. These curves indicate that an especially strong rise of T_2 occurs during the increase of z from 0 to $4 \cdot 10^{-3}$ deg^{-1} (in the limit as $z \to \infty$, $\Delta T_{2\,\max}$ approaches T_0 and $T_{2\,\min}$ approaches 0°K). The semiconducting materials used at present in industry (with $z = 2 \cdot 10^{-3}$ deg^{-1}) make it possible to achieve only a relatively small cooling effect: $\Delta T_{2\,\max} = 50\text{-}70°$. Future development of semiconducting materials with $z = 10^{-2}$ deg^{-1} or more should make it possible to achieve a greater cooling effect while still using low-grade heat sources. The absolute value of ΔT_2 may be considerably greater than the value of ΔT_1. For example, when $n = 10$, $l_1/l_2 = 0.5$, $T_0 = 323°K$, and $z = 10^{-2}$ deg^{-1} the curves of Fig. 4 give $(\Delta T_2)_{\max} = 150°$, while $(\Delta T_1)_{\max} = 50°$.

97

The problem of the consumption of liquid coolant is very important. The amount of heat transferred to the liquid in our device is $N = (Q_2 + Q_1)/Q_2$ times greater than in a thermoelectric refrigerator of the same heat-pumping rate but working from an external d-c source. However, if we use the "spent" liquid of the thermoelectric refrigerator as a coolant for a thermoelectric generator, then the connection of the generator to the refrigerator will produce only an increase (by a factor of N) of the temperature difference in this liquid between the input and output of the device without changing the consumption of the liquid. Such an increase of the temperature of the generator coolant is permissible because its harmful effect may be compensated by a corresponding increase of the temperature T_1. It is then possible to use a closed liquid circulation system and cooling radiators.

The principle of the simultaneous use of thermoelectric effects can serve as the basis of cooling devices (large as well as very small) operated by means of cheap low-grade heat sources: spent hot water from factories and thermal power stations, geothermal sources, solar energy, etc.

The authors are very grateful to I. L. Gerlovin for his interest in this work and discussions.

LITERATURE CITED

1. I. N. Pomazanov and P. L. Tikhomirov, Izvest. Akad. Nauk UzSSR, Ser. Fiz.-Mat. Nauk, No. 3, 1960.
2. I. N. Pomazanov and P. L. Tikhomirov, Kholodil'naya Tekh. No. 4, 1961.
3. A. F. Ioffe, Proc. Tenth International Refrigeration Congress, Copenhagen, 1959.
4. S. I. Angello, Recent progress in thermoelectricity, Elec. Eng. 79(5), 1960.
5. L. S. Stil'bans, Semiconductor Thermoelectric Refrigerators, Leningrad Dom Nauch.-Tekh. Propagandy, 1957.

DESIGN OF THERMOPILES UNDER NON-STEADY-STATE CONDITIONS

A. G. Shcherbina

The operation of semiconductor refrigerators and heaters in the initial period before steady-state conditions are reached is frequently of considerable practical interest.

Most often, an allowance for the specific features of the thermal processes under non-steady-state conditions is necessary when the operating conditions of a thermoelectric device require a knowledge of the time needed to reach steady-state conditions or when the desired temperature of a cooled (or heated) object should be reached in a definite time.

The methods of solving such problems, especially when high accuracy is needed, require mathematical functions to describe a wide range of pehomena during the transient period. However, a rigorous solution of this problem has not yet been obtained in the general form. Moreover, a rigorous solution of even the simplest special cases [1, 2] shows that the required mathematical functions are very complex and not very suitable for engineering applications. This is a consequence of the fact that in a thermocouple the temperature variation is not simply exponential. If the operating conditions of a thermoelement are restricted and the true thermal processes are replaced by acceptable models, the process of establishing a temperature may be described by a simple exponential law which does not differ greatly from the true dependence. For this purpose, it is sufficient to restrict the working current to the maximum current $I_m = vT_c/R$ and to regard the heat supply (or removal) to the semiconducting material and the heated (or cooled) object as a volume process. In our analysis, we shall consider a single thermoelement (thermocouple).

We shall begin with the simplest operating conditions. When the temperature of the hot junctions may be regarded as constant and equal to the temperature of the ambient medium, the heat balance of the cooled object (semiconductor) may be written in the following way:

$$Q_0 d\tau = \left(\frac{c_p \cdot g}{\varphi} + cG \right) dT_c + \alpha f (T_h - T_c) d\tau, \tag{1}$$

where Q_0 is the heat-pumping rate of a single thermoelement; τ is the time; c_p is the specific heat of the semiconducting material; c is the specific heat of the bridge plate at the cold junction; g is the weight of the semiconducting material in a single thermoelement; G is the weight of the bridge plate at the cold junction; α is the heat-transfer coefficient; f is the surface taking part in heat exchange, per thermoelement[*]; T_h and T_c are the temperatures of the hot and cold junctions, respectively; φ is a coefficient which allows for the "effective mass" of the semiconducting material of the thermoelement taking part in the cooling process.

According to the theory of thermoelectricity [3], Q_0 is, in general, determined by the steady-state condition at the cold junction and has the following form:

$$Q_0 = vIT_c - 0.5I^2R - \frac{v^2(T_h - T_c)}{Rz}, \tag{2}$$

---

[*] When the cold junction is not finned, the overall heat transfer is affected strongly by the side surfaces of the thermoelement branches next to the cold junction. In that case, it is necessary to include about a third of the side surfaces of the thermoelement in the effective heat-transfer surface.

where I is the working current; $R = 2h/\sigma s$ is the resistance of the thermoelement; v is the average thermoelectric power of the thermoelement in the working range of temperatures; z is the figure of merit representing the efficiency of the semiconducting material; h is the height of the thermoelement; s is the cross section of the thermoelement branches; σ is the conductivity of the semiconducting material.

Before integrating Eq. (1), it is necessary to find whether the coefficient φ depends on the cold-junction temperature T_c. For this purpose, we shall write the coefficient φ in the following way:

$$\varphi = \frac{T_h - T_c}{T_a^0 - T_a}, \tag{3}$$

where T_a^0 and T_a are the average temperatures along the thermoelement height under steady-state conditions with temperature drops at the junctions equal to zero and $T_h - T_c$, respectively.

The value of T_a will be determined from the formula

$$T_a = \frac{\int_0^h T\,dx}{h}. \tag{4}$$

We can easily show that the temperature distribution along the thermoelement height has the following form:

$$T = T_h - \frac{T_h - T_c}{h} \cdot x + \frac{2I^2 zh}{\sigma^2 s^2 v^2}(hx - x^2); \tag{5}$$

here x is the running coordinate (the origin of coordinates is at the hot junction). Integrating Eq. (4) with allowance for Eq. (5), we obtain

$$T_a = \frac{T_h + T_c}{2} + \frac{I^2 h^3 z}{3v^2 \sigma^2 s^2}. \tag{6}$$

The value of T_a^0 is obtained from Eq. (6) by substituting $T_h = T_c$:

$$T_a^0 = T_h + \frac{I^2 h^3 z}{3v^2 \sigma^2 s^2}. \tag{7}$$

Substituting Eqs. (6) and (7) into Eq. (3), we find that the coefficient φ depends neither on the working current nor on the temperature difference at the junctions, and it is equal to 2.

Integrating Eq. (1) with allowance for Eq. (2), we obtain the following time dependence for the cooling process:

$$\tau = \tau_0 \ln \frac{vIT_h - 0.5I^2 R}{Q_0^\tau - \alpha f(T_h - T_c)}; \tag{8}$$

here Q_0^τ is the heat-pumping rate of the thermoelement at the end of the period considered, i.e., at the time

$$\tau_0 = \frac{0.5 c_p g + cG}{\dfrac{v^2}{Rz} + vI + \alpha f}. \tag{}$$

From Eq. (8), we obtain the final expression for the reduction of the cold-junction temperature as a function of time τ:

$$T_c = T_h - \Delta T_{st}\left(1 - e^{-\frac{\tau}{\tau_0}}\right), \tag{9}$$

where ΔT_{st} is the temperature difference between the thermoelement junctions under steady-state conditions:

$$\Delta T_{st} = \frac{vIT_h - 0.5I^2R}{\frac{v^2}{Rz} + vI + \alpha f} \ .$$ (10)

We note that, when $\alpha = 0$ and $I_m = vT_c/R$, we have

$$\Delta T_{st} = \Delta T_{max} = \frac{zT_c^2}{2} \ .$$

To determine the time required for the temperature difference between the thermoelement junctions to reach a given value ΔT, we can easily obtain the following equation from Eq. (9):

$$\tau = \tau_0 \ln \frac{\Delta T_{st}}{\Delta T_{st} - \Delta T} \ .$$ (11)

Since τ_0 has the dimensions of time, this quantity can be regarded as a time constant, the physical meaning of which can be easily found from Eq. (9) by substituting in it $\tau = \tau_0$.

Carrying out appropriate calculations, we find that $\Delta T = 0.632 \Delta T_{st}$, which shows that the time constant τ_0 is the time during which the temperature difference across the thermoelement reaches about two-thirds the temperature difference under steady-state conditions.

It may be shown also that, under conditions of perfect thermal insulation of the cold junction during the time interval equal to the time constant τ_0 and under ideal operating conditions, the thermoelement reaches the same temperature difference between the junctions which it would reach under real steady-state conditions, i.e., ΔT_{st}.

Allowance for the Temperature Variations of the Bodies Surrounding the Working Junction

We have assumed so far that the temperature of the medium surrounding the cold junction is constant and the only variable parameter is the cold-junction temperature. Thermoelements work rarely under such conditions. Frequently, a thermopile cools a thermally insulated volume in which there are various devices, some of which evolve heat at their own specific rate.

Let us consider the case when the hot junctions are in contact with a moving liquid, the temperature of which will be assumed to be constant and equal to the temperature of the surrounding medium. In this case, the temperature drop between the hot junction and the moving liquid (water) is usually small and may be neglected by assuming that the hot-junction temperature is equal to the temperature of the liquid. If these conditions are satisfied then the thermal balance for the thermoelement and the cooled bodies may be written in the following way:

$$Q_0 d\tau = (0.5 c_p g + \sum p_i c_i G_i) dT_c + \alpha f (T_a^0 - T_c) d\tau,$$ (12)

where

$$T_a^0 = T_0^0 - \frac{Q_0 - \omega_0}{k_0 f_0'} = T_h - \frac{Q_0 - \omega_0}{k_0 f_0'} ; \quad p_i = \frac{T_h - T_{m_i}}{T_h - T_c} ,$$

T_a^0 is the temperature of the medium around the cold junctions (fins) under steady-state conditions; α is the heat-transfer coefficient with reference to the temperature difference between the base of the cold fins and the heat-transfer medium; f is the surface through which heat is exchanged between the cold junction of one thermoelement and the ambient medium (air); T_0^0 is the temperature of the medium surrounding the cooled chamber; T_{m_i} are the average temper-

101

atures of the cooled bodies under steady-state conditions; k_0 is the heat-transfer coefficient of the cooled-volume enclosure; f_0' is the surface of the thermal insulation of the cooled volume, per thermoelement; ω_0 is the heat evolution inside the cooled volume, per thermoelement; c_i are the specific heats of the cooled bodies; G_i are the weights of the cooled bodies, per thermoelement. The remaining notation has been explained above.

The solution of Eq. (12) leads to the same equations (9) and (11), but now the values of ΔT_{st} and τ_0 are different:

$$\Delta T_{st} = \frac{b\left(vIT_h - 0.5I^2R\right) - \dfrac{af}{k_0 f_0'}\omega_0}{b\left(\dfrac{v^2}{Rz} + vl\right) + af} \tag{13}$$

and

$$\tau_0 = \frac{0.5c_p g + \sum p_i c_i G_i}{b\left(\dfrac{v^2}{Rz} + vT\right) + af}, \tag{14}$$

where

$$b = \left(1 + \frac{af}{k_0 f_0'}\right).$$

In the calculation of ΔT_{st} and τ_0, it is necessary to select the coefficients p_i as follows. Their values represent the ratios of the degree of cooling of the bodies in the enclosure to the temperature drop at the cold junction during the time necessary to reach the steady state. As a rule, the values of p_i do not exceed 0.7-0.9.

In a real device, the temperature difference between the cooled body and the cold junction of the thermoelement under steady-state conditions can usually be estimated with an acceptable accuracy and from this difference it is not difficult to determine the values of p_i. Under similar operating conditions, the heat balance for a thermoelement used as a heater ($T_c = const = T_0^h$) is given initially by

$$Q_h d\tau = \left(0.5c_p g + \sum p_i' c_i' G_i'\right) dT_h + \alpha' f'\left(T_h - T_{op}^\tau\right)d\tau; \tag{15}$$

$$Q_h = vIT_h + 0.5I^2R - \frac{v^2\left(T_h - T_c\right)}{Rz},$$

$$p_i' = \frac{T_{mi}' - T_0^h}{T_h - T_0^h}, \quad T_a^h = T_0^h + \frac{Q_h + \omega_h}{k_h f_h'} = T_c + \frac{Q_h + \omega_h}{k_h f_h'},$$

where T_0^h is the temperature of the medium surrounding the heated volume (chamber); T_a^h is the temperature of the medium surrounding the hot junctions (fins) under steady-state conditions; k_h is the heat-transfer coefficient of the heated-volume enclosure; f_h' is the surface of the thermal insulation of the heated volume, per thermoelement; ω_h is the heat evolution inside the heated volume (chamber), per thermoelement; c_i' are the specific heats of the heated bodies; G_i' are the weights of the heated bodies, per thermoelement; T_{mi}' are the average temperatures of the heated bodies under steady-state conditions; α' is the heat-transfer coefficient with reference to the temperature difference between the base of the hot fins and the heat-transfer medium; f' is the surface through which heat is exchanged between the hot junction of one thermoelement and the ambient medium. Solution of Eq. (15) leads to the following dependences:

$$T_h = T_c + \Delta T_{st}'\left(1 - e^{\frac{-\tau}{\tau_0}}\right), \tag{16}$$

$$\tau = \tau_0' \ln \frac{\Delta T_{st}'}{\Delta T_{st}' - \Delta T}, \qquad (17)$$

where

$$\Delta T_{st}' = \frac{b'(vIT_c + 0.5I^2R) + \frac{a'f'\omega_h}{k_h f_h'}}{b'\left(\frac{v^2}{R_s} - vI\right) + a'f'},$$

$$\tau_0' = \frac{0.5c_p q + \sum p_i' c_i' G_i'}{b'\left(\frac{v^2}{R_s} - vI\right) + a'f'},$$

$$b' = \left(1 + \frac{a'f'}{k_h f_h'}\right).$$

Derivation of Formulas for Calculating the Conditions When the Hot and Cold Junction Temperatures Are Not Fixed

The initial equations (12) and (15) can be written as follows:

$$Q_0 d\tau = M dT_c + U(T_a^0 - T_c) d\tau, \qquad (18)$$

$$Q_h d\tau = N dT_h + Y(T_h - T_a^h) d\tau; \qquad (19)$$

here $M = 0.5c_p g + \sum p_i c_i G_i$, $N = 0.5c_p g + \sum p_i' c_i' G_i'$, where U and Y are the products of the co-efficient of heat transfer between a cold (hot) junction and the heat-transfer medium multiplied by the heat-transfer area of one junction. The remaining notation is the same as before.

Carrying out the necessary transformations, we obtain from Eqs. (18) and (19) the following system of linear equations with constant coefficients:

$$\frac{dT_c}{d\tau} = \gamma T_c - \rho T_h - \nu, \qquad (20)$$

$$\frac{dT_h}{d\tau} = p T_c - m T_h + n, \qquad (21)$$

where

$$\rho = \frac{\rho'}{M}; \quad p = \frac{p'}{N}; \quad \gamma = \frac{\gamma'}{M}; \quad m = \frac{m'}{N}; \quad \nu = \frac{\nu'}{M}; \quad n = \frac{n'}{N}; \quad \rho' = \frac{v^2}{R_z}\left(1 + \frac{U}{k_0 f_0'}\right);$$

$$p' = \frac{v^2}{R_z}\left(1 + \frac{Y}{k_h f_h'}\right); \quad \gamma' = \left(1 + \frac{U}{k_0 f_0'}\right)\left(\frac{v^2}{R_s} + vI\right) + U;$$

$$m' = \left(1 + \frac{Y}{k_h f_h'}\right)\left(\frac{v^2}{R_z} - vI\right) + Y; \quad \nu' = \left(1 + \frac{U}{k_0 f_0'}\right)0.5I^2R + UT_0^0 + \frac{U\omega_0}{k_0 f_0'};$$

$$n' = \left(1 + \frac{Y}{k_h f_h'}\right)0.5I^2R + YT_0^r + \frac{Y\omega_h}{k_h f_h'}.$$

Differentiating Eq. (20), we obtain

$$\frac{d^2T_c}{d\tau^2} = \gamma \frac{dT_c}{d\tau} - \rho \frac{dT_h}{d\tau}. \qquad (22)$$

From Eq. (20) we find T_h in terms of τ, T_c, and $dT_h/d\tau$:

$$T_h = \frac{\gamma}{\rho}T_c - \frac{1}{\rho}\cdot\frac{dT_c}{d\tau} - \frac{\nu}{\rho}.$$

Substituting T_h into Eq. (21), we find $dT_h/d\tau$ in terms of the same quantities:

$$\frac{dT_h}{d\tau} = pT_c - \frac{m\gamma}{\rho} T_c + \frac{m}{\rho} \cdot \frac{dT_c}{d\tau} + \frac{\nu m}{\rho} + n. \tag{22a}$$

Then Eq. (22) with allowance for Eq. (22a) becomes

$$\frac{d^2T_c}{d\tau^2} + A\frac{dT_c}{d\tau} - BT_c = -D; \tag{23}$$

here $A = m - \gamma$; $B = m\gamma - p\rho$; $D = \nu m + n\rho$. The particular solution of Eq. (23) with $T_c = D/B$ gives the value of the cold-junction temperature under steady-state conditions. Similarly, we can find the expression for the hot junction:

$$\frac{d^2T_h}{d\tau^2} + A\frac{dT_h}{d\tau} - BT_h = -E, \tag{24}$$

where $E = n\gamma + \nu p$.

The particular solution of Eq. (24) with $T_h = E/B$ gives the hot-junction temperature under steady-state conditions.

The solution of Eqs. (23) and (24) is the sum of the special solution and the solution without the right-hand part:

$$T_c = C_1 e^{-|r_1|\tau} + \frac{D'}{B'}, \tag{25}$$

$$T_h = C_2 e^{-|r_2|\tau} + \frac{E'}{B'}, \tag{26}$$

where $|r_1|$ and $|r_2|$ are the absolute values of the positive and negative roots of the characteristic equation $r^2 + Ar - B = 0$; $D' = \nu'm' + n'\rho'$; $B' = m'\gamma' - \rho'p'$; $E' = n'\gamma' + \nu'p'$.

To determine the constants C_1 and C_2, we use the initial conditions

$$T_c(\tau=0) = T_0',$$

$$T_h(\tau=0) = T_0''.$$

Substituting $\tau = 0$ into Eqs. (25) and (26), we find that

$$C_1 = T_0' - \frac{D'}{B'}; \quad C_2 = T_0'' - \frac{E'}{B'}.$$

Thus the final expressions for T_h and T_c become:

$$T_c = \frac{D'}{B'} + \left(T_0' - \frac{D'}{B'}\right)e^{-|r_1|\tau}, \tag{27}$$

$$T_h = \frac{E'}{B'} - \left(\frac{E'}{B'} - T_0''\right)e^{-|r_2|\tau}. \tag{28}$$

From the expressions (27) and (28), we obtain the dependences for the determination for the time necessary for T_h and T_c to reach the required values:

$$\tau = \frac{1}{|r_1|} \ln \frac{T_0' - \frac{D'}{B'}}{T_c - \frac{D'}{B'}}; \tag{29}$$

$$\tau = \frac{1}{|r_2|} \ln \frac{\frac{E'}{B'} - T_0''}{\frac{E'}{B'} - T_h}.$$

LITERATURE CITED

1. L. S. Stil'bans and N. A. Fedorovich, Zhur. Tekh. Fiz. 28:490, 1958.
2. J. E. Parrot, Solid State Electronics, 1(2):135, 1960.
3. A. F. Ioffe, L. S. Stil'bans, E. K. Iordanishvili, and T. S. Stavitskaya, Thermoelectric Cooling, Izd. Akad. Nauk SSSR, Moscow-Leningrad, 1956.

INDEX